YELLOWSTONE

The aptly named Beartooth Mountains, jutting like a carnivore's jawline of rugged wildness into the sky north of Yellowstone National Park, mark a dramatic entrance to the most iconic ecosystem in the lower 48 states.

MICHAEL NICHOLS

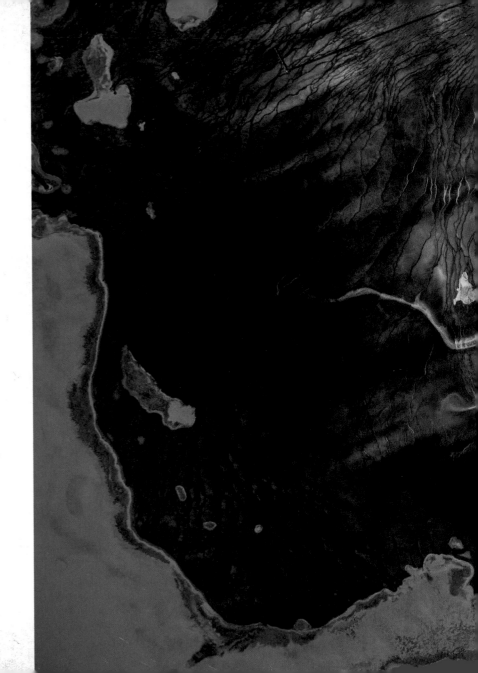

Grand Prismatic
Spring, richest in
color and mightiest
of Yellowstone's hot
springs, was first ref-
erenced by mountain
man Osborne Russell
in his 1839 journal.

MICHAEL NICHOLS

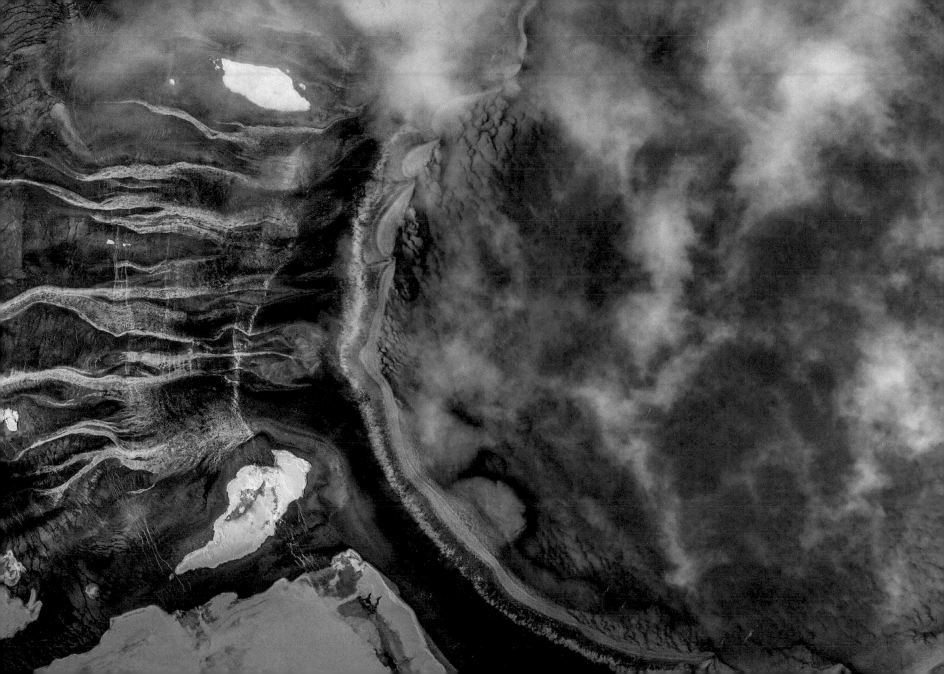

The American plains bison, once numbering in the tens of millions, began its journey back from the brink of extinction more than a century ago when just two dozen survivors were given refuge in Yellowstone National Park. Today some of their wild and free-ranging descendants find a home in Jackson Hole, Wyoming, like this herd roaming across the National Elk Refuge.

CHARLIE HAMILTON JAMES

Unforgettable even if viewed only once, the Teton Range rises as the famous centerpiece of Grand Teton National Park. As a new day dawns at Schwabacher's Landing, a bend of the Snake River renowned for its diversity of wildlife, the Tetons are indeed the epitome of purple mountains' majesty.

CHARLIE HAMILTON JAMES

A JOURNEY THROUGH AMERICA'S WILD HEART

YELLOWSTONE

DAVID QUAMMEN

NATIONAL GEOGRAPHIC

WASHINGTON, D.C.

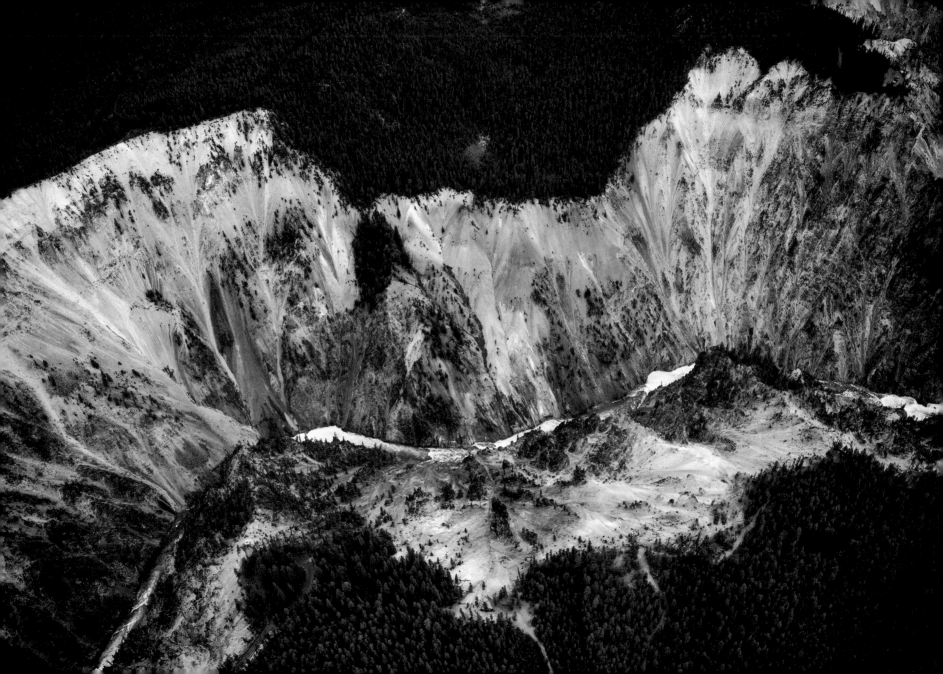

CONTENTS

(page 10)

Few animals in Greater Yellowstone rival the majesty of bull elk. These imposing monarchs, resplendent with their large antlers, rest in their Yellowstone day beds.

MICHAEL NICHOLS

The gnawed-out maw of the Grand Canyon of the Yellowstone is a scene of sublime beauty from both air and ground level. Roaring through the chasm, the untamed Yellowstone River has carved a tumbling, wending course into walls made of rhyolite, deposited by ancient volcanic eruptions.

MICHAEL NICHOLS

THE GREATER YELLOWSTONE ECOSYSTEM

The idea of a Greater Yellowstone Ecosystem, which gained traction in the 1980s, marked a giant leap forward in ecological thinking. Biological processes within Yellowstone Park extend far beyond its borders, and activities in one place can have huge implications for the area next to it. The construct is now helping managers everywhere understand that a healthy, well-functioning bioregion—Greater Yellowstone encompasses 22.6 million acres—is worth far more than the sum of its parts.

AREA ENLARGED
UNITED STATES

Elevation
10,000 ft
8,000
6,000
4,000

20 mi

MONTANA
WYOMING

Billings
Lovell
Red Lodge
Powell
Cody
Lander
Pavillion
Ocean Lake

Clark's Fork Yellowstone

BEARTOOTH MOUNTAINS
Granite Peak 12,799 ft 3,901 m
Cooke City
Barronette Peak
Amphitheater Mt.
Cache Creek
Soda Butte Cr.
Fall Rose Creek
Mt. Everts
Tower

A B S A R O K A

Buffalo Bill Reservoir
North Fork Shoshone

R A N G E
Deer Creek Pass
Open Cr.
Deer Cr.
Thorofare Cr.

WIND RIVER RANGE
Gannett Peak 13,804 ft 4,207 m
Fremont Lake
Green
Pinedale

Big Timber

Livingston

GALLATIN RANGE

Gardiner
Mammoth
Mt. Washburn
Blacktail Deer Plateau
Mirror Plateau
Lamar Valley
Pelican Creek
Yellowstone Caldera
Mt. Langford
Two Ocean Plateau
Trident Plateau
Thorofare Plateau

Y E L L O W S T O N E

M O U N T A I N S

E C O S Y S T E M

Bozeman
Belgrade

GREATER

Mammoth Hot Springs
Ran Sepulcher
Monument Mt.
Norris Geyser Basin
Steamboat Geyser
Grand Canyon of the Yellowstone
Hayden Valley
Sulphur Valley
Yellowstone Lake
Sylvan Pass
Flat Mt.
Shoshone Lake
YELLOWSTONE N.P.
JOHN D. ROCKEFELLER JR. MEMORIAL PARKWAY
Jackson Lake
GRAND TETON NATIONAL PARK
NATIONAL ELK REFUGE
GROS VENTRE RANGE

Big Sky
Gallatin
MADISON
RANGE
Grand Prismatic Spring
Old Faithful
Madison Plateau
Mesa Falls

Jackson
Jackson Hole
TETON RANGE

WYOMING RANGE
SALT RIVER RANGE

Ennis Lake
Hebgen Lake
Henrys Lake
West Yellowstone
RED ROCK LAKES NATIONAL WILDLIFE REFUGE

Driggs

WYOMING
IDAHO

Virginia City

Madison

Island Park Reservoir
MESA FALLS SCENIC BYWAY

Ashton
St. Anthony

Palisades Reservoir

Paris

ROCKY

MONT.
IDA.
Lima Reservoir

Dubois

Rexburg

Preston

Boulder
Butte

Jefferson

Beaverhead

Mud Lake

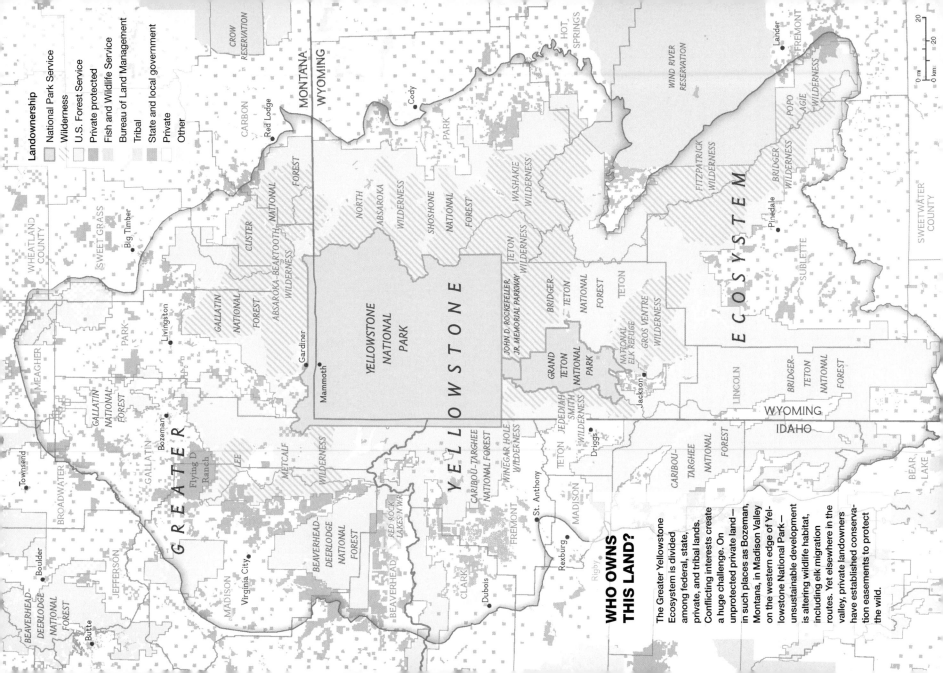

Landownership

- National Park Service
- Wilderness
- U.S. Forest Service
- Private protected
- Fish and Wildlife Service
- Bureau of Land Management
- Tribal
- State and local government
- Private
- Other

WHO OWNS THIS LAND?

The Greater Yellowstone Ecosystem is divided among federal, state, private, and tribal lands. Conflicting interests create a huge challenge. On unprotected private land—in such places as Bozeman, Montana, in Madison Valley on the western edge of Yellowstone National Park—unsustainable development is altering wildlife habitat, including elk migration routes. Yet elsewhere in the valley, private landowners have established conservation easements to protect the wild.

MONTANA
WYOMING

WYOMING
IDAHO

GREATER

YELLOWSTONE

ECOSYSTEM

YELLOWSTONE NATIONAL PARK

GRAND TETON NATIONAL PARK

CROW RESERVATION

WIND RIVER RESERVATION

SHOSHONE NATIONAL FOREST

NORTH ABSAROKA WILDERNESS

WASHAKIE WILDERNESS

TETON WILDERNESS

BRIDGER-TETON NATIONAL FOREST

GROS VENTRE WILDERNESS

NATIONAL ELK REFUGE

JOHN D. ROCKEFELLER, JR. MEMORIAL PARKWAY

FITZPATRICK WILDERNESS

POPO AGIE WILDERNESS

BRIDGER WILDERNESS

BRIDGER-TETON NATIONAL FOREST

JEDEDIAH SMITH WILDERNESS

TETON WILDERNESS

WINEGAR HOLE WILDERNESS

CARIBOU-TARGHEE NATIONAL FOREST

CARIBOU-TARGHEE NATIONAL FOREST

RED ROCK LAKES N.W.R.

BEAVERHEAD-DEERLODGE NATIONAL FOREST

GALLATIN NATIONAL FOREST

LEE METCALF WILDERNESS

GALLATIN NATIONAL FOREST

CUSTER NATIONAL FOREST

ABSAROKA-BEARTOOTH WILDERNESS

CUSTER NATIONAL FOREST

BEAVERHEAD-DEERLODGE NATIONAL FOREST

SWEET GRASS

WHEATLAND COUNTY

CARBON

PARK

CARBON

MEAGHER

BROADWATER

JEFFERSON

MADISON

FREMONT

CLARK

MADISON

LINCOLN

SUBLETTE

FREMONT

SWEETWATER COUNTY

BEAR LAKE

Lander

Cody

Red Lodge

Big Timber

Livingston

Gardiner

Mammoth

Bozeman

Flying D Ranch

Townsend

Boulder

Butte

Virginia City

Dubois

St. Anthony

Rexburg

Rigby

Driggs

Jackson

Pinedale

0 mi   20
0 km   20

PART ONE

The Paradox of the Cultivated Wild

The wild life of America exists in the consciousness of the people as a vital part of their national heritage.

—GEORGE M. WRIGHT, JOSEPH S. DIXON, AND BEN H. THOMPSON,
Fauna of the National Parks of the United States, 1933

A lone bison trudges along the shore of ice-covered Mary's Bay on the north side of Yellowstone Lake. Having endured another winter, this hungry young bull, a member of the historic Pelican herd, will soon be rewarded with a panorama of green grass.

MICHAEL NICHOLS

Yellowstone has long been a complicated idea in the American mind, as well as a wondrous reality on the ground.

It's a name—"yellow stone" in English, *roche jaune* to the French explorers, *Mi tse a-da-zi* to the Hidatsa people who had been in the region centuries longer—of uncertain provenance but probably inspired by yellowish sandstone bluffs along a tranquil stretch of river in what is now eastern Montana. Hence the Yellowstone River, still the longest undammed waterway in the contiguous United States. But the name traveled upstream on that river as the waters flowed down. It came to signify a very different place—a place different from

Sidewinding across the Midway Geyser Basin, the Firehole River has an ambiance unmatched by any other water corridor in Yellowstone National Park.

anywhere—hundreds of miles southwest of the yellowish stone bluffs, in the mountains where that river begins. This was a more severe landscape, a high-altitude wilderness, uplifted above the plains and rimmed by sharp peaks—a bizarre and spectacular redoubt that the Hidatsa themselves seldom if ever visited.

Some native peoples did go up there, distant ancestors of those now known as Sheep Eater, Bannock, Crow, and others, moving on and off that highland hideaway as their nomadism led them in search of food and furs and seasonally comfortable living. Because the elevation made for especially brutal winters, this land wasn't fought over, seized, and settled during the early waves of Euro-American invasion.

A few mountain men and fur trappers had seen bits of it, including John Colter and Jim Bridger, who told tales. Then, in 1872, through an improbable chain of connivances, as frontier and native cultures gave way to white immigrants and towns and railroads, one vast rectangle of that mysterious terrain became Yellowstone National Park.

It wasn't just America's first national park. It was (if you take the words strictly, to mean a park for the nation, designated by national law) the world's first. It represented a radical idea—far more radical than its conceivers and its promoters recognized at the time—that was destined to survive dire challenges, grow in force and complexity, and become only more valuable, though also more fraught, with passing time. The novelist Wallace Stegner famously called national parks "the best idea we ever had," a phrase echoed by Ken Burns and Dayton Duncan in their television series *The National Parks: America's Best Idea*. With all due respect to them, it's worth noting that this park concept was confused, inchoate, in some ways cynical at the start. Officials of the Northern Pacific Railroad saw profits to be made; politicians were assured that there would be no public costs, only benefits. Still, if not our very best idea (Jefferson's Declaration of Independence had articulated a few valuable notions, after all), it was a good idea that has gotten better, a big idea that has gotten bigger over time.

Now we have many parks, of many sorts and sizes, spread from Kobuk Valley in Alaska to Everglades in Florida, and from Haleakala in Hawaii to Acadia in Maine. Yellowstone is still special. The story of Yellowstone National Park plus its adjacent lands—and the adjacent lands *must* be part of the story—shows us that this "best" idea had mixed origins and that it has always been a work in progress, initially vague, unforeseeably complex, continually evolving, more contentious today than ever.

Yet it *is* a great idea, the idea of national parks, and Yellowstone will remain its greatest embodiment, not just for America but also for the world, if we can only agree what we want the place to be, and to do, and to mean.

Uphill From All Directions

The first thing you notice about Yellowstone National Park, if you're paying attention, is that the approach is uphill. It's uphill from all directions. Coming through the north gate, in the town of Gardiner, Montana, you'll follow a two-lane blacktop that climbs upstream along a little river and then rises away, traversing a mountainside, switchback by switchback, to the park headquarters up at Mammoth, Wyoming. Stay alert for the bighorn sheep on the cliffs above this road, which may send a rock crashing onto your car. If you drive west out of Cody, Wyoming, you'll ascend for 50 miles between old, russet volcanic bluffs toward the park's east entrance just short of Sylvan Pass. Brake for grizzlies. On the west side, you might take the picturesque route from Ashton, Idaho, passing suddenly from potato fields into conifer forest as you divert along the Mesa Falls Scenic Byway. This back road will carry you uphill past an overlook, where the Henry's Fork of the Snake River drops 114 feet over the upper falls, then another 65 feet over the lower, two modest increments signaling that the river itself is tumbling off what geologists call the Yellowstone Plateau.

Atop the plateau lies the park, with an average elevation of 8,000 feet. Once you're up there, large stretches of forest and grassland will seem almost flat, at least compared with your steep approach. You can drive for miles along a gently undulant road, through dense stands of lodgepole pine, across ground that appears cold and static. Don't be fooled.

There's a dramatic geological reason for the height of the Yellowstone Plateau. Directly beneath it burns a vast volcanic hot spot, a gigantic channel through Earth's mantle and crust, through which heat rises and melts the rocks, creating a great ascending plume. That thermal torrent comprises two magma chambers of partly molten rock, one atop the other, bulging the land surface into an enormous pustule. Around the bulge, like disorderly ramparts, loom mountains that are older and higher—most notably the Tetons, the Absarokas, the Gallatin Range. Amid the plateau itself, geologists have traced the evidence of three huge ovals, calderas, representing the scars left by three stupendous explosions over the past 2.1 million years. Those explosions, and the volcanic forces that powered them, have earned Yellowstone's hot spot the label "supervolcano." Ordinary volcanoes generally occur along the edges of tectonic plates; supervolcanoes blaze directly through those plates, like a stationary torch burning blisters through a sliding sheet of steel. And the Yellowstone torch, feeding heat toward preposterous eruptions, is likely the largest beneath any continent on Earth.

"It all starts with heat," according to Robert B. Smith of the University of Utah, who has studied Yellowstone's geology for over five decades. While the North American plate has drifted southwestward over the mantle plume during the past 16 million years, the hot spot has left its marks northeastward in a series of volcanic centers, 500 miles long, from what we now call southeastern Oregon and across Idaho to its current location. The most recent of the three giant upheavals at Yellowstone occurred about 640,000 years ago, spewing 240 cubic miles of debris into the atmosphere and leaving a rimmed crater that now encompasses Old Faithful, the Hayden Valley, and half of Yellowstone Lake. That one is known as the Yellowstone Caldera. (There have also been many smaller volcanic eruptions in the millennia since.) The implications of these geological facts are indirect but fateful. After the ash settled and the land cooled, the Yellowstone Plateau remained a site of extraordinary volcanic activity, its relatively thin earthly crust floating above the hot spot's upper magma chamber, which heats the subterranean waters that emanate as geysers and fumaroles and mud pots and colorful hot springs, all penetrating the surface like whistles on a great calliope. Gradually the plateau's forests regrew, its animal populations recolonized. Meanwhile, a combination of gouging and splitting forces, including ice, flowing water, and geological faulting, opened a channel along which the Yellowstone River carved its own Grand Canyon (big and impressive, though not nearly as vast as Arizona's), thundering over a pair of spectacular drops that make Mesa Falls look sedate and puny.

The geysers and the Grand Canyon of the Yellowstone would eventually play lead roles in making this a place with statutory protections. In the years 1869–1871, long after Colter and Bridger had spread steamy rumors, three different expeditions of more citified white men, along with some military personnel, visited the area and took appreciative note of its scenic wonders, in particular the geysers and the canyon. One of those men, Nathaniel P. Langford, has been described (by Aubrey L. Haines, author of a two-volume history of Yellowstone) as "a sickly St. Paul bank clerk" who made his timely exit from Minnesota after a family-owned bank failed. Langford imagined better opportunities, less embarrassing circumstances, and came west. While playing a catalytic role in the 1870 Yellowstone expedition, led by Henry Washburn, Langford was a paid publicist for the Northern Pacific Railroad. Another member of that expedition, Walter Trumbull, noted afterward in a magazine article that the plateau seemed promising as sheep pasture, but he predicted: "When, however, by means of the Northern Pacific Railroad, the falls of the Yellowstone and the geyser basin are rendered easy of access, probably no portion of America will be more popular as a watering place or summer resort." Langford and his cronies saw that such popularity would mean money in the tills of the Northern Pacific, and of whoever else got a piece of the action, selling rail tickets, filling hotels.

The 1871 expedition, led by Ferdinand V. Hayden, head of the U.S. Geological Survey of the Territories, was more official—supported by a modest congressional appropriation, given in the spirit of national expansion and inventory—and included the photographer William H. Jackson and the painter Thomas Moran. Jackson's photos and Moran's canvasses subsequently helped people back east (most crucially, those in Congress) see and imagine Yellowstone. Moran created one especially thunderous painting in 1872, 7 feet by 12, "The Grand Canyon of the Yellowstone." An agent for the Northern Pacific then planted a suggestion that lawmakers protect "the Great Geyser Basin" as a public park. Hayden seized that idea and, along with Langford and other minions of the railroad, lobbied for it, as delineated in a bill encompassing not just the geyser basins but also the Grand Canyon of the Yellowstone, Mammoth Hot Springs, Yellowstone Lake, the Lamar Valley, and other terrain, altogether a rectangle of some 2.1 million acres.

The Yosemite Valley in California, which had earlier been granted to that state for protection as a state park, served as a rough precedent; Niagara Falls back in New York, on the other hand, stood as a negative paradigm. Niagara was infamous to anyone who cared about America's natural majesties, because private operators there had bought up the overlooks and blocked the views, turning that spectacle into a commercial peep show. Yellowstone, as a great public attraction promising to bring visitors and money westward, would be different.

Legendary Yellowstone superintendent Horace Albright dines with three park bears in 1924. Feeding bruins was encouraged by park officials early on as tourist entertainment.

Congress embraced what the Northern Pacific, Ferdinand Hayden, and others had proposed, and on March 1, 1872, President Ulysses S. Grant, compliant but no great advocate of scenic protection himself, signed a bill creating Yellowstone National Park. That act, not surprisingly for its time, ignored any prior claims by the Sheep Eater or other native groups. It specified "a public park or pleasuring-ground for the benefit and enjoyment of the people," meaning implicitly non-Indian people. Within this park, "wanton destruction of the fish and game," whatever "wanton" might mean, as well as commercial exploitation of such game, was prohibited. The boundaries were rectilinear, although ecology isn't. It was almost as though a vast wilderness had been neatly framed.

This Massive Slaughter

We rode along gravel in the ranger's big white truck, turning left at a sign that said AUTHORIZED VEHICLES ONLY. Ahead of us stood a large barn, a small office, and a series of fenced pastures, within which dozens of bison munched contentedly on hay that had been spread for them on dry February ground. This bit of Yellowstone, known as the Stephens Creek Capture Facility, remains unseen by park visitors, and the bison too were oblivious to their situation. The fur atop their huge heads was curly, dark, and stiff; their withers were thickly robed; their hind legs seemed ridiculously delicate; and their languid brown eyes attended only to the hay. They had no idea what was coming. Just beyond the barn was a complex of small corrals, holding pens, chutes, and gates, from which these "excess" bison from the park would be loaded, next morning, to be trucked away for slaughter at meat plants in Montana.

Yellowstone National Park played a crucial role, more than a century ago, in the rescue of the American bison from extinction. But now there are too many bison for the amount of habitat that society allows them. For "society," read: the Montana Department of Livestock. Bison aren't permitted to migrate seasonally across the park boundary into Montana when winter hunger drives them down, because some of them carry an infectious bacterial disease, brucellosis, which is considered a threat to the economics of cattle ranching. Montana designates Yellowstone bison "a species requiring disease control" and doesn't welcome them as free-ranging wildlife. Elk carry brucellosis too, but elk are allowed transboundary migration, an inconsistency Montana policy declines to address. The meat, the heads, and the hides of these doomed bison at Stephens Creek would go to individual Indian tribes and the InterTribal Buffalo Council, of Rapid City, South Dakota, salvaging a compensatory good from a bad and controversial situation.

As I watched, a blood sample was drawn from each bison, with the powerful animal pinched helplessly in a squeeze chute, the veterinary phlebotomist working deftly to avoid a broken arm in the scuffle, the blood tubes going for immediate processing nearby in a small, wood-heated cabin. There they were spun down on a centrifuge, then run through an electronic gizmo to detect antibodies against brucellosis. Most of them, I was told, would test positive for this disease—an affliction that's alien to the Yellowstone region, brought in by European cattle and transmitted to bison, which now are condemned for the slim possibility that they will transmit it back. I stood in the cabin, watching the blood work with two concerned men: Rick Wallen, a large, soft-spoken bison biologist with a Smith Brothers beard, and P. J. White, his boss. White serves as chief of Wildlife Resources for Yellowstone. He's a stocky fellow with a gray mustache and an aquiline nose, the burly candor of a high school football coach, and a sharp brain. He expressed his dissatisfaction with the illogic of Montana's brucellosis-defense campaign: "I'm not willing to decimate our bison herd when nothing's being done on elk." Willing or not, though, White was obliged to watch that decimation happen.

Bison containment is just one more in a long series of dilemmas that have faced Yellowstone since its establishment. At the outset, the park was an orphan idea with no clarity of purpose, no staff, no budget. Congress seemed to lose interest as soon as the ink of Grant's signature dried. Yellowstone became a disaster zone, neglected and abused, for more than a decade. Nathaniel Langford, the failed bank clerk and railroad publicist, served as its first superintendent, at zero salary, and during his five years in the post he barely earned that, revisiting the park only two or three times. Market hunters established themselves brazenly in the park, killing elk, bison, bighorn sheep, and other ungulates in industrial quantities. By one account, a pair called the Bottler brothers shot about 2,000 elk near Mammoth Hot Springs in early 1875, generally taking only the tongue and the

hide from each animal, leaving the carcasses to rot or be scavenged. That account doesn't say how many grizzly bears the Bottlers killed over those carcasses, for convenience or profit, but undoubtedly the elk meat was a dangerous attractant that brought bears near guns. An elk hide was worth six to eight dollars, serious money, and a man might kill 25 to 50 elk in a day. "There was this massive slaughter that occurred here, from 1871 through at least 1881," according to Lee Whittlesey, currently Yellowstone's historian. Antlers littered the hillsides. Wagon tourists came and went unsupervised, at low numbers but with relatively high impact, some of them vandalizing geyser cones, carving their names on the scenery, killing a trumpeter swan or other wildlife for the hell of it. Ungulate populations fell, and then the carnage gradually petered out, Whittlesey told me, "until the army arrived here in 1886."

As an act of desperation, in the absence of any congressional appropriation for managing Yellowstone or any trained body of park police to enforce its rules, the secretary of the interior in 1886 asked the U.S. Army to take over. And with that event, an unlikely hero enters the story: Gen. Philip H. Sheridan.

Philip Sheridan is best known, and most infamously remembered, as a ruthless cavalry leader under Grant during the Civil War and, later, as commander of the horrific military campaigns against the Plains Indians. His troops attacked the Cheyenne, the Kiowa, the Comanche, the Utes, the Sioux, killing many and driving survivors onto reservations. He advocated exterminating the buffalo as a means of crushing tribal cultures and resistance. But after he visited

Yellowstone in 1882, a more appealing side of Sheridan's character emerged. In this very different context, he deplored the slaughter of "our noble game," evidently even the bison, and offered troops to prevent it. He was also appalled that a commercial monopoly on visitor services had been granted to the so-called Yellowstone Park Improvement Company, a new entity closely allied with the Northern Pacific Railroad. "I regretted exceedingly to learn," he reported to Washington, "that the national park had been rented out to private parties." And he made one radically percipient observation: Congress had made the park too small.

Returning to Washington, Sheridan led a campaign by sportsmen and sympathetic lawmakers to extend Yellowstone's boundaries by 40 miles along the east side and 10 miles along the south. That would have increased the park area by 2.1 million acres, almost doubling its size. More crucially, it would have added adjacent lowlands, to which elk and other ungulates migrate in winter.

Carried into Congress by Senator George G. Vest of Missouri, the Sheridan proposal failed. The boundaries stood. Those boundaries were later tweaked, in the 1920s and 1930s, adding some smallish areas and removing one, to reflect stream drainages rather than abstract lineation. But the need for winter range by big herbivores—especially elk and bison—remained a festering problem, and it still festers today.

The first guardians of Yellowstone were Army infantry officers, not park rangers. Here visiting soldiers affiliated with the U.S. Army Bicycle Corps pose on Minerva Terrace at Mammoth Hot Springs in 1896.

The Bear That Eats Flowers

Lines on a map give an overview, but the best overview comes from 300 feet airborne above the landscape itself. So on a clear summer morning, I met Roger Stradley at an airport near Bozeman, Montana, for an aerial tour of Yellowstone in his yellow 1956 Piper Super Cub. He issued me a flight suit and a helmet, then showed me how to insert myself into the rear seat, a cramped space directly behind the pilot's. I've been in kayaks with more legroom. "You don't get *into* a Super Cub," he said. "You put it *on*." Small and light, with a high-lift wing and a strong engine, the Super Cub is still favored by bush pilots such as Stradley for its capacity to land and take off from short strips, and to fly slowly, riding the thermals like a condor.

Flying slow and low is especially useful for surveying wildlife, and Stradley is legendary among the biologists of Yellowstone Park who rely on overflights for their work, counting and radio-tracking bears, wolves, elk, and other wide-ranging animals. On the day I rode along, he was a robust, cheerful 76-year-old and had flown in the mountains for 62 years, logging a mere 70,000 hours. Before that, from the age of about six, he had sat on the lap of his dad, an old-school mountain pilot, who let little Roger control the stick, though he couldn't reach the pedals. His younger brother Dave got the same training. They would bank and dip and try to hold a good line as their father laughed, keeping the plane stable with his feet. "We drove him all over the sky." Dave retired from their Gallatin Flying Service a few

FLIGHT OVER YELLOWSTONE

Beginning and ending in Bozeman, Montana, pilot Roger Stradley provided writer David Quammen with an aerial tour of the Greater Yellowstone Ecosystem. Their course took them east through wolf-rich Lamar Valley and then looped south to the isolated Thorofare region before heading west over geyser-studded basins.

me through my helmet headphones to a wolf den, signaled by bare dirt and dig marks beneath the exposed roots of a large Douglas fir. Stradley saw, though I couldn't, a single black wolf lying there in the shade. That animal would be part of the old Eight Mile Pack, he said, which had since fissioned, as wolf packs do when they get too large. Doug Smith, Yellowstone's chief wolf biologist, often flew in the seat where I was sitting, and when Smith wasn't along, Stradley helped track the wolves himself. He had been gazing down at them, after all, for 20 years, since their reintroduction to the park in 1995. And now he carried his own radio receiver, near at hand in the tiny cockpit, so he could tell one pack from another by their signals.

We traced the Yellowstone River upstream past Tower Fall, catching a flash of rainbow through the mist, and then crossed the Antelope Creek basin toward Mount Washburn, passing over uplands patched with brown, dead trees from a recent fire. Good bear habitat, Stradley said. Elk to eat, plus lots of mushrooms, morels, sprouting up from the burn. "You don't want to go in there and pick them, because you're going to have competition with the grizzlies." Duly noted.

We swung back toward the Lamar Valley, broad and grassy between sage-covered hillsides, punctuated with glacial erratics (large boulders from elsewhere left behind by the moving ice), and sparsely visited by tourists because, here in the park's northeastern corner, it lies off the great double loop road that carries people to less subtle glories: the geyser basins, Yellowstone Lake, and the Grand Canyon. The Lamar is a magical place on several counts, one being the curious relationship between glacial erratics and Douglas

years ago, but Roger was now facing knee surgery (plus cancer therapy, which he mentioned lightly) so that he could continue.

We rose up over Bozeman, crossed the Gallatin Range, stopped at the Gardiner airstrip for a pee, then rose again and circled into the park. We flew east along the gently sloped north side of Mount Everts, named for Truman Everts, an unlucky man who went missing from the 1870 expedition and survived a month of lonely misery before being rescued. We crossed the Blacktail Deer Plateau, a loaf of raised terrain within the vaster Yellowstone Plateau, where Stradley alerted fir trees in the lower valley: Wherever you see one of those cabin-size boulders amid the sage flats, you see also a single Doug fir risen snugly beside it, bark to stone, evidently because the harsh conditions—drought, cold, wind, or whatever—make it impossible for a cone to open, a seed to germinate, or a sapling to survive, anywhere *but* in the lee of a boulder. "Nurse rock" is one label for the stony half of this partnership. I've mused over the relationship across almost 40 years of visits to the Lamar, never found a scientific study to explain it, and now from overhead I mused again: the love story of the boulder and the tree.

We circled above Rose Creek, site of one of several acclimation pens from which the 1995 wolves were released. They went forth and multiplied, as the scientists had predicted they would, thriving on a diet of overabundant elk. We could see yellow flowers peppered across the meadows of grass and sage—the rich yellows of balsamroot, the paler yellows of biscuitroot, which signal a tuber that bears eat. Lacking geysers and a great canyon, the Lamar Valley isn't for everyone, thank goodness, but to some eyes it's the most exquisite corner of Yellowstone.

Back in the 1880s, certain schemers lobbied Congress to obtain a right-of-way through this valley, which would have severed it from park protection, so that the misnamed Yellowstone Park Improvement Company (YPIC), with the Northern Pacific behind it, could run a railroad line from Gardiner to the mining town of Cooke City, just outside the northeast entrance. Cutting that swath down the valley, bringing through steam-powered transport, would have made it a bustling commercial corridor, not a haven for grizzlies and wolves and people content to gaze on big trees smooching boulders. The schemers included even Yellowstone's superintendent of the moment, a lackey of the YPIC named Robert Emmett Carpenter. But word leaked, opposition rose, Superintendent Carpenter was canned, and it didn't happen. The present two-lane blacktop through the Lamar, along which small clusters of wolf-watchers crouch behind spotting scopes in all seasons, their SUVs parked at a few turnouts, is a benign alternative.

We looped eastward above Soda Butte Creek, a Lamar tributary, and toward two distant, high teeth of the Absaroka Range—Index Peak and Pilot Peak, beyond the park boundary in the Shoshone National Forest. Barronette Peak, on our left, still held a large snowfield, which no doubt suited the mountain goats that favor its eastern face. Stradley banked steeply, pulling us into a grand turn amid a huge glacial cirque he called "the Amphitheater," U-shaped and high-walled, spooned out by ancient ice from what the maps label Amphitheater Mountain. We circled a second time, gaining enough altitude to climb over its east rim. The Amphitheater, as Stradley noted, is an underappreciated majesty. "Nobody sees it. Nobody knows it's here." The circuit was thrilling, but my stomach settled more comfortably as we leveled out over Cache Creek and then crossed the Mirror Plateau.

Pelican Creek spills off the Mirror Plateau toward Yellowstone Lake, draining a wide backcountry basin that's excellent grizzly bear habitat in early summer—so excellent that it's closed to hikers until mid-July in consideration of the danger. We saw a few elk, and some bison, amid grassy expanses dotted with small, dark tarns. Along the edge of the valley, large swaths of bare, dead timber still stood as scorched monuments to the great fires of 1988, which had tormented park managers and challenged hundreds of firefighters through that long, dreadful summer. Critics of fire policy found themselves unable, at the time, to imagine that Yellowstone could ever be lush and verdant again, but the fire ecologists have been proved right: Flame comes, flame goes; fire is part of the natural cycle of temperate forests. That the 1988 fires burned hotter and more extensively than lightning-caused fires normally would—that was unfortunate, yes, but not ecologically catastrophic, and represented consequences of a century of fire-suppression policy (with Smokey the Bear as emblem), inordinate fuel buildup, extended drought, and high winds at inopportune moments, not willful mismanagement, as the critics claimed. From the air, the mosaic of scars reflecting those interconnected conflagrations of 1988 now lingered more conspicuously in some parts of the park—such as here—than in others. The fire scars, and their greening over time, echoed a broader Yellowstone theme: disturbance and recovery.

As we approached the lower reaches of Pelican Creek, where it meanders through willowy flats, we spotted a bison carcass in the water, with a grizzly on it, feeding. Forty feet away, watching patiently, waiting their turn, sat two black wolves. "That's a big bear," Stradley said. The wolves may have been thinking the same thing.

The bison carcass wasn't noticeable when he flew through here yesterday, Stradley added. It must have been dead in the shallows, a winterkill, kept chilled by the water, until the bear found it and opened it up. That would release the smell, bringing other meat eaters. "I don't know where the ravens are. They should be here already." Almost a mile on, we passed above another big grizzly, a dark boar, walking upwind toward the dead bison. Claims would be adjudicated by strength, ferocity, and persistence; little would go to waste. Ravens, when they did arrive, would dodge in and out boldly, pecking away bits while the grizzlies or the wolves were too occupied to object.

We flew down the eastern shore of Yellowstone Lake into another roadless area, protected on one side by water and another by the Absaroka Range, here including Mount Langford, named for the man who shilled Yellowstone on behalf of the Northern Pacific. Most visitors to the park remain within a few hundred yards of their cars, and therefore never see such bits of the wondrous, difficult Yellowstone backcountry. Near the southern tip of the lake's Southeast Arm is the delta of the upper Yellowstone River, a broad bottomland of willows, grasses, and shrubs in five shades of green. The lake receives water from little creeks on all sides, but its largest input is this stream, draining north from a vast wilderness outside the park's southeastern corner. Today the upper river was brownish olive, reflecting fast July runoff from melting snow, though I noticed one oxbow pond, its water a deep tranquil blue.

We flew upstream in stately progress, so low and slow that it felt like we were riding a kite. We crossed the park boundary, an invisible

World-renowned Jackson Hole mountaineer and adventure photographer Jimmy Chin captures the Tetons during a sunrise takeoff.

JIMMY CHIN

line, and turned east to follow a tributary called Thorofare Creek toward the rocky headwaters where the Yellowstone River has one of its sources. "This down here, where the Thorofare is," Stradley said, "is the most remote spot in the continental United States." He meant the lower 48, not a continent that includes Alaska, but I got the point. "Thirty miles from anything."

Circling back into the park, we crossed Two Ocean Plateau, so named because its gentle hump divides rivulets running east to the Yellowstone from rivulets running west to the Snake River. Whereas the Yellowstone carries its waters on a great loop to the Missouri River and then eastward to the Mississippi and south to the Gulf of Mexico, the Snake drains westward to the Columbia River and the Pacific—another indicator of how the Yellowstone Plateau punctuates America.

After three hours of flying we needed to point ourselves back toward the airport, but first we swung over the Norris Geyser Basin, which includes Steamboat Geyser, the world's tallest though far from its most faithful. Steamboat erupts at irregular intervals and to heights of almost 400 feet, where it could have fogged our windshield if the timing had been just right; but Steamboat held off, and instead we had a clear view of all the little hot springs and ponds dotting the area in shades of aquamarine, orange, yellow, and chartreuse.

Ahead of us then was the Gallatin Range, not as lofty and jagged as the Absarokas, though high enough to retain snow patches and cornices at this season. On one of those white patches we glimpsed a cluster of dark figures and, swooping closer, made out their forms: seven wolves plus a grizzly bear, sharing a little snowfield at uneasy proximity and an elevation of 9,000 feet. That seemed peculiar, so Stradley banked hard to make another pass.

Reaching over his shoulder, he handed me the radio receiver to free his hands. He pulled up his camera to get a photo of the animals as we circled back. Now he was flying the plane with his knees. "I don't know why they're up so high," he said. "What's the attraction?"

We couldn't see, we couldn't guess, and before I knew it we were two mountains along, looking down on another bear, a huge one. This grizzly was on a flat above Fan Creek, munching contentedly amid a patch of yellow balsamroot and other vegetation. Again we circled. Stradley seemed puzzled, and the natural historian in him wanted clarification. Balsamroot is not usually mentioned as a grizzly food, although the bear's dietary choices in Yellowstone are formidably diverse, according to a recent authoritative paper; the list includes 266 kinds of plant, animal, and fungus, ranging from bison flesh to morels, from western waterweed to moose, and from chipmunks to 25 kinds of grass. Was the bear grazing? Didn't seem to be. Digging up tubers? No. "Might be eating the flowers," Stradley said. And it was possible: Kerry Gunther, the park's chief bear biologist, first author on that authoritative paper, later told me that balsamroot flowers have indeed been found in grizzly scat.

So that's what I took away, my last vivid impression from our eagle-view tour: a humongous grizzly bear alone on a hill above the Gallatin River, eating flowers.

The Good, the Bad, and the Cute

Wolves don't eat flowers. Mountain lions don't eat grass. But the grizzly bear is an omnivore, not a pure carnivore, and for that reason among others it has always occupied a unique status in the Yellowstone ecosystem and the American mind.

During the late 19th and the early 20th century, predatory animals suffered ruthless persecution within Yellowstone Park, both as a matter of neglect (while poaching was rampant) and as a matter of ill-conceived policy. The very notion that the park should protect wildlife as well as geysers and canyons was an afterthought, and that afterthought initially applied only to the "good" creatures, the game animals that hunters prized, the trout that fishermen prized, the benign herbivores that visitors could comfortably admire, such as elk and deer, pronghorn and moose, bison and bighorn sheep. Bison were a special concern because the grotesque excesses of market hunting, mindless thrill shooting from railroad cars, and the United States Army's anti-Indian campaign after the Civil War, under Sheridan, had nearly driven them extinct. Some of that bison massacre occurred even in Yellowstone. The army itself would be given custodianship of the park in 1886, as a desperation measure against rampant poaching, but it wasn't until passage of the Lacey Act in 1894—an act "to protect the birds and animals in Yellowstone National Park, and to punish crimes in said park"—that Yellowstone's caretakers had authority to arrest and prosecute offenders. The army would continue in its role until 1918, when administrators and wardens from the newly formed National Park Service (NPS) arrived. Although the Lacey Act helped both the army and then the NPS to protect Yellowstone wildlife, it came almost too late for the bison.

By 1901 only a few hundred bison remained in America, about two dozen of which had found refuge in the Pelican Valley. Yellowstone officials managed to save those few, bred them with captive bison brought from ranches elsewhere, and eventually created a bison-ranching operation in the Lamar Valley—straightforwardly called the Buffalo Ranch—with its animals ranging free during summer and herded back in autumn, to spend winter in corrals eating hay. The logic and merits of the Buffalo Ranch were questionable, from an ecological perspective, but as a reaction to the scare over bison extinction it was understandable. The ranch operated until 1952.

Horace Albright, superintendent of the park from 1919 to 1929, worried that America's elk might also face extinction, because of uncontrolled slaughter outside the park and the harsh winters they faced on the Yellowstone Plateau. Albright is remembered as one of Yellowstone's great figures, a beloved and heroic administrator, but in fact his legacy is ambivalent. He had served as a young assistant to Stephen Mather, the man chiefly responsible for creating the National Park Service, and he shared Mather's commitment to raising Yellowstone's value, and that of a system of national parks generally, by increasing tourism. Large, visible herds of elk represented popular attractions, so Albright wanted them—even at the expense of other native Yellowstone animals. He instituted a program of

feeding hay to elk during winter, hoping to keep them from migrating out of the park and into danger from hunters, and he encouraged his rangers to kill predators, especially wolves, mountain lions, and coyotes. Sportfishing was another visitor draw, so white pelicans, gorgeous big birds but nefarious trout eaters, were suppressed at their breeding colonies. One colony was on Molly Island, a remote site in the Southeast Arm of Yellowstone Lake, where few tourists ever saw it. Beginning in 1924, under Albright's administration, park personnel with help from the U.S. Fish Commission crushed pelican eggs and killed chicks on Molly Island. An assistant chief ranger reported about 200 eggs destroyed in 1926 and 83 young pelicans killed in 1927; in 1928, he added proudly if ungrammatically, "we killed 131 every young pelican on the island, that year nothing escaped."

Persecution of the "bad" animals in Yellowstone was an old project, though, dating well back before Albright. Predators had been shot, trapped, and poisoned since the 1870s. One superintendent even encouraged commercial trappers to kill beaver by the hundreds, so that they wouldn't build dams and flood his park. Otters were classified as predatory, damning them also to suppression, and for a while there was a fatwa against skunks. During the army years, both noncommissioned officers and civilian scouts were "authorized and directed to kill mountain lions, coyotes, and timber wolves," by order of the secretary

of the interior. Wolf killing ended only when the wolves were all gone, extirpated not just from Yellowstone (by around 1930) but throughout the American West. Poisoning and shooting of coyotes continued until about 1935. But bears were different.

Bears were omnivorous and, as some people saw them, cute. They were also smart and opportunistic. Beginning as early as 1883, they adjusted to feeding on food refuse from garbage dumps near the park hotels, and that behavior made them easily visible and therefore a popular tourist attraction. They also learned to accept handouts from passing visitors, a trend that started in the stagecoach era and continued after private automobiles were allowed into the park, beginning in 1915. Albright encouraged the handouts game, leading people to think that bears—even grizzlies—were companionable, benign, and feckless. Near the hotels at Old Faithful, Lake, and Canyon, the dumps became theaters where tourists sat on bleacher seats to watch the "bear show" on summer evenings. The historian James A. Pritchard, documenting this enterprise (and Albright's view of it), wrote that "if two thousand people could crowd into an amphitheater" and see a number of grizzlies feeding on garbage, "the scale of the attraction simply attested to Yellowstone's preeminent place as a showpiece of nature."

Not just during Albright's tenure but for 80 years, Yellowstone's grizzlies and black bears consumed human garbage in vast quantities, coming to depend on it unwholesomely, with the blessings of the park managers and to the amusement of the visiting public. "One of the duties of the National Park Service," Albright himself wrote, after

Two American classics from the past converge in Yellowstone's Hayden Valley— one whose time has passed, the other experiencing a renaissance of appreciation, nominated by Congress to be the country's official land mammal.

succeeding Mather as head of the NPS in 1929, "is to present wild life 'as a spectacle.' This can only be accomplished where game is abundant and where it is tame."

But the grizzlies of Yellowstone were never tame.

Diorama in Blood

On August 7, 2015, a ranger in Yellowstone found the chewed-on body of a man, dead and cold, near a hiking trail not far from Lake Village, one of the park's busiest visitor developments. Although there were no witnesses, gruesome evidence made plain that this man had suffered an encounter with a grizzly bear. Tracks at the scene showed that the bear was an adult female, traveling with at least one young cub. The victim was soon identified as Lance Crosby, 63 years old, from Billings, Montana. He had worked seasonally as a nurse at a medical clinic in the park and been reported missing by co-workers that morning.

Investigation revealed that Crosby was hiking alone on the previous day, August 6, without bear spray (the recommended repellent), and ran afoul of this female grizzly—with her two cubs, not just one. The sow, after killing and partially eating him (not necessarily in that order) and allowing the cubs to eat too, cached his remains beneath dirt and pine duff, as grizzlies do when they intend to reclaim a piece of meat. Once trapped and persuasively linked to Crosby by DNA evidence, she was given a sedative and an anesthetic and then executed, on grounds

that an adult grizzly who has eaten human flesh and cached a body is too dangerous to be spared, even if the fatal encounter wasn't her fault. "We are deeply saddened by this tragedy, and our hearts go out to the family and friends of the victim," said Superintendent Dan Wenk, a reasonable man whose heart presumably went out to the bear too.

Lance Crosby's death was just the seventh bear-caused human fatality in Yellowstone National Park over the past hundred years. The next most recent occurred in 2011, when two people died in separate events within that summer, possibly killed by a single female grizzly. The 2011 bear had been spared by Wenk's decision (with advice from bear managers) after the first killing, of a man named Brian Matayoshi, on grounds that she was defending her cubs and the attack on Matayoshi wasn't predatory. Because the Matayoshi killing occurred on the Wapiti Lake trail, she became known as the Wapiti sow. Later she turned up in the vicinity of the second victim, a man named John Wallace, who died eight miles away in what might or might not have been a predatory attack. Wallace's body, like Lance Crosby's, had been partially eaten and then cached. Physical evidence didn't prove that the Wapiti sow had killed him, but it strongly suggested that she had at least fed on the body; and so she was put down. The sorry events of 2011, and that well-meant choice to give the Wapiti sow a reprieve after Matayoshi's death, help explain the decision to condemn the 2015 sow after one incident.

Grizzly bears, clearly, can be dangerous animals. But the danger they represent, to human visitors and workers within Yellowstone National Park, should be seen in perspective: Within the past 144 years since Yellowstone was established, more people have died there of drowning, and of scalding in thermal pools, and of suicide than have been killed by bears. Almost as many people have died from lightning strikes. Two people have been killed by bison. The real lesson inherent in the death of Lance Crosby, and in the equally regrettable death of the unfortunate bear that killed him, is a reminder of something too easily forgotten: Yellowstone is a wild place. Sort of.

It's a wild place that we have embraced, surrounded, encompassed, riddled with roads and hotels and souvenir shops, but not tamed, entirely. It's filled with wonders of nature—fierce animals, deep canyons, scalding waters—that are magnificent to behold but fretful to engage. Most of us, when we visit Yellowstone, see it as if through a Plexiglas window. We gaze from our cars at a roadside bear, we stand at an overlook above a great river, and we stroll boardwalks amid the geyser basins, experiencing the place as a diorama. We remain safe and dry. Our shoes don't get muddy with sulfurous gunk. But the Plexiglas window doesn't exist, and the diorama is real. It's painted in blood—the blood of many wild creatures, dying violently in the natural course of relations with one another, predator and prey, and occasionally with the blood also of humans. Walk just 200 yards off the road into a forested gully or sagebrush flat, and you had better be carrying, as Lance Crosby wasn't, a canister of bear spray. Your park entrance receipt won't protect you. Your senior pass (I have one too) definitely won't protect you. You can be killed and eaten. But if you are, despite your own choices, there may be retribution.

This is the paradox of Yellowstone, and of most other national parks in America that we have added since: wilderness contained, nature under management, wild animals obliged to abide by human rules. It's the paradox of the cultivated wild. At a national park in Africa—Serengeti in Tanzania, for instance, or Masai Mara in Kenya, Kruger in South Africa—you wouldn't face such ambiguity. You would view the dangerous beasts, the lions and elephants and leopards and buffalo, from the safety of your Land Rover or your safari van, seldom if ever strolling through their habitat on foot. But in America we've chosen to do things differently—and Yellowstone, because it came first, because it contains grizzly bears, because millions of people visit each year, is where the paradox takes most acute form.

Yellowstone nowadays is more than a park. It's an iconic name known throughout the world that stands for a wild idea in the American mind, a wild place in the American West, a wild heart in the American breast, still beating after 144 years. It's also the eponym of a great ecosystem, the biggest and richest complex of mostly untamed landscape and wildlife within the lower 48 states.

Question: Can we hope to preserve, in the midst of modern America, *any* such remnant of our continent's primordial condition, *any* such sample of true wildness—a gloriously inhospitable landscape, full of predators and prey, in which nature is still allowed to be red in tooth and claw? Can that sort of enclave be reconciled with human demands and human convenience? Can we carry it from our rough national past into the crowded, paved, unimaginably technological future? Time alone, and our choices, will tell. But if the answer is yes, the answer is Yellowstone. ∎

(preceding pages)
A liquid jewel rimmed on its eastern flank by the Absaroka Mountains, Yellowstone Lake is North America's largest lake above 7,000 feet, with 110 miles of shoreline. Artifacts confirm a human connection going back millennia.

RAUL TOUZON

Dwellers of the higher mountain ramparts, bighorn sheep with their distinctive curled headgear were hunted by an offshoot band of Shoshone Indians known as the Tukudeka, or "Sheep Eaters," in and around Yellowstone.

BARRETT HEDGES

The deep guttural bellows of bull bison sparring for dominance during mating season have been compared to the roars of African lions.

(opposite)
A nursery band of bison cows and their newborn calves find good grazing on a slope near Specimen Ridge in Yellowstone's Lamar Valley.

As deep snows melt away above the tree line, grizzlies head higher to eat on plants and, in some places, gorge themselves on army cutworm moths, which migrate hundreds of miles from farm fields on the plains to drink nectar from mountain wildflowers.

It's not identified on any tourist maps, but the discovery of this "bear bathtub"—a spring-fed watering hole where grizzlies and black bears come to drink, bathe, and frolic—has yielded fascinating insights into bear behavior thanks to on-site remote cameras.

MICHAEL NICHOLS WITH RONAN DONOVAN AND THE NATIONAL PARK SERVICE

(opposite)
A grizzly bear mama floats in the cool waters of the bear bathtub, her twin cubs waiting to take the plunge.

MICHAEL NICHOLS WITH RONAN DONOVAN AND THE NATIONAL PARK SERVICE

Francine Spang-Willis

EDUCATOR, MONTANA STATE UNIVERSITY

"I am a descendant of Chief Dull Knife and Pawnee Woman of the Northern Cheyenne Nation. I am also a descendant of a pioneer family. The work I do in archaeology and education helps me to help others have a deeper understanding of the American Indian connection to the landscape and our shared American history. There are numerous sites in the Yellowstone area that remind us that our ancestors lived and interacted with every inch of this sacred landscape for thousands of years. Artificial boundaries may exist on the land, but we still have a deep connection to all of it. We still live and interact with it."

A member of the Northern Cheyenne Tribe, anthropologist Francine Spang-Willis surveys the view from the Bridger Mountains, north-west of Yellowstone, a range ironically named after a white fur trapper and not one of the Native Americans who lived there long before his 1823 arrival.

ERIKA LARSEN

The Fire Below

More than a third of Yellowstone sits within the caldera of a giant ancient yet still active volcano.

It takes a geologist to reveal the hidden reality—to summon, for instance, the ghosts of the jagged, 12,000-foot Teton-like peaks you would have seen if you'd stood in the park three million years ago. "They're gone because an eruption 2.1 million years ago blew them to smithereens," says Robert B. Smith of the University of Utah. Another massive eruption followed 640,000 years ago. It's not over, Smith says: The Yellowstone supervolcano is "living, breathing, shaking, baking."

The massive, superheated plume of rock under Yellowstone fuels the Earth's largest collection of hydrothermal features. Eruptions of this supervolcano—so named for the violence and size of its explosions—expel so much material that the crust caves in, creating craterlike depressions called calderas. Evidence of past eruptions shifts with the Earth's crust, a migrating testament to the still active forces below.

Seismometers in the park record 1,000 to 3,000 earthquakes a year. Most are too small to be felt by humans (though a magnitude 7.3 temblor in 1959 killed 28 people). The giant caldera that Smith and other geologists have mapped—a crater left by ancient eruptions that pumped out epic volumes of lava and ash—occupies more than a third of the park's 2.2 million acres. But most visitors will drive right over the caldera's rim and never know it. "I've long believed that when visitors come into the park, they could be greeted with a sign that says, 'Welcome to Yellowstone—you're now entering a volcano,'" Smith says. "Yellowstone wouldn't be Yellowstone if not for this reality."

Streaks of fluorescent orange appear to be erupting over the brim of Grand Prismatic Spring, but the colors come from pigmented bacteria— also called thermophiles—that live in hot water flowing out of the ground.

GORDON WILTSIE

Ethereal drama in twilight: As a thunderstorm blows in, Fan Geyser and Mortar Geyser stand side by side on the banks of the Firehole River, erupting together in concert.

MICHAEL NICHOLS

(opposite)
Great Fountain, located just off Firehole Lake Drive, remains a crowd pleaser, erupting roughly once or twice a day. The largest gusher in the Lower Geyser Basin, it was first documented by the Cook-Folsom-Peterson expedition in 1869, three years before Yellowstone became America's first national park.

MICHAEL NICHOLS

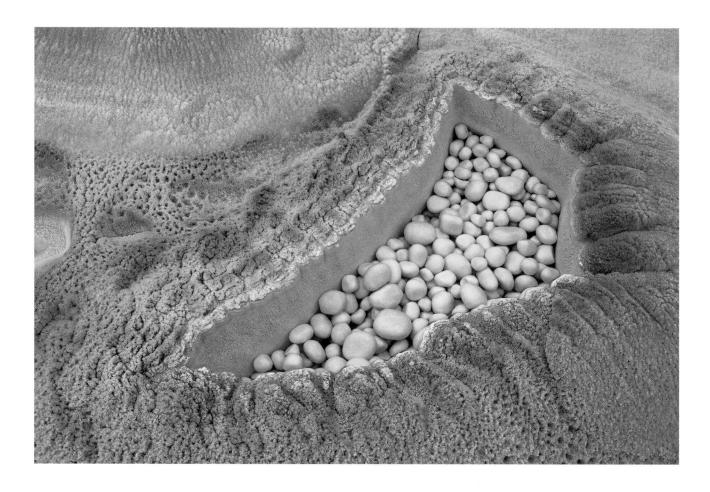

Yellowstone's innumerable hot springs often hold treasure, like this pool of sinter—hardened mineral deposits—polished through the rock tumbler of time and now naturally collected in a crusty crevice.

MICHAEL NICHOLS

(opposite)
A lone female hiker vanishes into the steam clouds billowing from Tardy Geyser. Like all Yellowstone geysers, it's born of snow or rainwater that has seeped deep into the ground, met superheated rock, and been blasted back out through a narrow blowhole.

MICHAEL NICHOLS

(following pages)

They might look like watercolor paintings portraying flaring tendrils of the sun, but the dazzling spectra of Grand Prismatic Spring (left) and Gentian Pool (right) actually reflect a palette of microscopic life— organisms specially adapted to live amid scalding water. The green is the chlorophyll they use to absorb sunlight. To protect it from human impacts, Gentian has no trail around its perimeter.

MICHAEL NICHOLS

Norris Geyser Basin is said to resemble a barren moonscape. As geysers erupt and fumaroles puff columns of sulfurous steam into the sky, the landscape feels like that of another planet.

MICHAEL NICHOLS

The Paradox of the Cultivated Wild 57

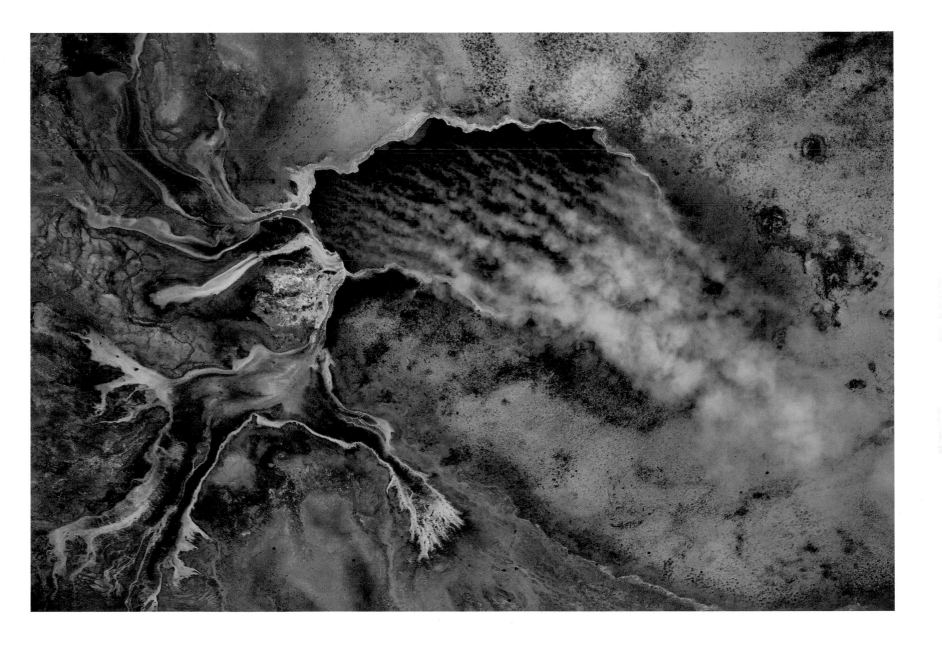

Visiting the Park

The motto Congress gave Yellowstone—"For the benefit and enjoyment of the people"—may sound straightforward . . . but it isn't.

Some call Yellowstone "the People's Park," and the numbers confirm the nickname. In 2015 the number of visitors to Yellowstone, including repeats, exceeded four million—a record number. Though the top destination remains Old Faithful, a recent study suggests visitors would be willing to pay significantly more to enter the park if they could count on seeing not just a geyser but also a bear along the road. Traffic jams generated by bear sightings can sometimes stretch for miles. All those visitors, to Grand Teton National Park as well as Yellowstone, pump an estimated billion dollars into the local economy.

Yellowstone superintendent Dan Wenk worries that the crowds, increasingly international, need to understand more about such matters as safe interaction with wildlife. And he worries about the park's future. On the one hand, Wenk wonders whether millennials see how important it is to preserve the park. On the other, he also worries that too many people come to the park already. It's an old worry: In 1972 a *National Geographic* article commemorating the park's 100th anniversary was titled "The Pitfalls of Success." Nearly twice as many visitors come to the park now as did then. The answer, Wenk thinks, isn't more hotels, roads, and parking lots. If anything, the human footprint needs to shrink to protect the wildlife the humans come to see. "What might work for Disneyland," Wenk says, "is not the antidote for Yellowstone."

The most famous geyser in the world, Old Faithful, so named for its cycle of predictable eruptions, has been one of Yellowstone's main attractions since the park was created in 1872. In earlier and less enlightened times, tourists were allowed to stroll literally to the gusher's edge.

UNDERWOOD & UNDERWOOD

(opposite)
Created to help spur ridership on the Northern Pacific Railroad line, Yellowstone advertising has for decades promised a destination where tourists could commune with nature, wildlife, and frontier traditions.

NATIONAL PARK SERVICE

1933

1910

1921

1923

1927

1910

At a prominent scenic overlook in Grand Teton National Park, a visitor records one view of the Tetons for posterity—with the real peaks rising in the background.

CHARLIE HAMILTON JAMES

Near West Thumb, off the southwestern corner of Yellowstone Lake, travelers gaze into the depths of an aquamarine hot cauldron bubbling with boiling water cooked by magma heat miles below.

ERIC KRUSZEWSKI

(opposite)
No summer pilgrimage to Yellowstone has been complete without a stop at Old Faithful Geyser. Legendary for its semireliable punctuality, the natural wonder erupts every 65 to 90 minutes and shoots up to a height of more than 180 feet.

MICHAEL NICHOLS

In the big open expanses of Midway Geyser Basin, humans walking the boardwalk around Excelsior Geyser are dwarfed by nature's scale. In contrast to urban environments where most people live, here there's a tone of humility in a landscape known to many as a wonderland.

The Paradox of the Cultivated Wild 65

Powerful, unpredict-
able, and potentially
dangerous, a bull elk
is swarmed by human
admirers along the
Grand Loop Road in
the center of Yellow-
stone. Park regula-
tions require people
stay 25 yards away
from wildlife, a rule
sometimes ignored,
often at visitors' peril.

ROBB KENDRICK

(opposite)
**The intersection of
human spaces and
wildness is what
makes navigating
through Yellowstone
unforgettable. Here,
a cow elk inspects
a pair of youthful
explorers on the
boardwalks at Mam-
moth Hot Springs.**

MICHAEL NICHOLS

First an outpost for cowboys in the late 19th century, Jackson Hole has evolved into a mecca for outdoor recreationists. Here, friends ride a chairlift to the top of Snow King Mountain, with the Tetons rising in the distance.

CHARLIE HAMILTON JAMES

(opposite)
The swim team from Santiago High School in Corona, California, braves Boiling River, a natural hot tub formed where Yellowstone's hot springs flow into the Gardiner River. Such springs can be perilously hot. More Yellowstone visitors have died by being scalded than by being mauled by bears.

MICHAEL NICHOLS

Like Old Faithful, Grand Geyser—the biggest geyser in Yellowstone—resides in the Upper Geyser Basin. Its eruptions are not as frequent, though: Grand spouts about three times a day. Its shooting stream can climb to 200 feet.

(opposite)

At a curio shop in Jackson, Wyoming, visitors posed with stuffed animals, including a brown bear—a Kodiak from Alaska, not a local grizzly. The desire to touch the quintessence of wildness, preferably without threat to life and limb, endures in many of us.

DAVID GUTTENFELDER

The presence of black bears in Yellowstone semitamed on human food was common into the early 1970s, when this photo was taken. Soon after, park policy shifted in an effort to wean begging bears off garbage and table scraps.

JONATHAN BLAIR

Dan Wenk

SUPERINTENDENT, YELLOWSTONE NATIONAL PARK

"All citizens of America, whether they realize it or not, are stewards of Yellowstone. It belongs to all of us … I believe we are rapidly coming to a point where one of two things is going to happen. Either we as a society agree to limit the number of visitors in order to protect resources that are incredibly sensitive to disturbance or we allow the number to grow unchecked—knowing that we are diminishing, perhaps irreparably, the very things that attract people worldwide to this one-of-a-kind national park."

After four decades with the National Park Service, Yellowstone superintendent Dan Wenk, the top protector in charge, intends to end his career here. "Few can ever say they got to look at the well-being of the first national park in the world," he says. "This, to me, represents the summit."

ERIKA LARSEN

PART TWO

The Return of the Wild

The air is electric and full of ozone, healing, reviving, exhilarating, kept pure by frost and fire, while the scenery is wild enough to awaken the dead.

—JOHN MUIR, 1901

A picture of placid serenity, a bull moose wades the Buffalo Fork of the Snake River in Grand Teton National Park. Habitat loss, predation by wolves and grizzly bears, diseases, drought, and hunting—all have caused a decline in the moose population throughout the Greater Yellowstone Ecosystem.

CHARLIE HAMILTON JAMES

What It's All About

On a cold morning in December at the Gardiner airstrip, I buckled into the back of a cherry-red Hughes 500D helicopter beside Doug Smith,

chief biologist for the Yellowstone Wolf Project. Seconds later the chopper levitated and then plunged toward the Yellowstone River, under the touch of Jim Pope, a wildlife-capture pilot with a sense of aerobatic panache. Pope leveled us off and then climbed again, sweeping south into the park, across the foothills, up over Sepulcher Mountain. Releasing a death grip on part of the cabin, I double-checked my seatbelt, since the helicopter was wearing no doors.

Frosty-faced after ascending a steep snow slope at Eagle Pass, a remote corner of the Greater Yellowstone Ecosystem located near Yellowstone's southeastern border, a grizzly triggers a remote-controlled camera trap.

Freezing wind ripped through our bubble as the treetops flashed by, 200 feet below. Then we set down gently on a clear patch of snow behind Sepulcher, where Pope's crew—a pair of "muggers," whose job was to fire a charge-propelled net, jump out, and tranquilize captured animals—had already immobilized two wolves.

The return of the wolf to Yellowstone is one of America's great conservation success stories. During 1995 and 1996, almost seven decades after eradication, 31 wolves from western Canada were released from acclimation pens in the Lamar Valley. They took hold of the new landscape, they bred, they proliferated, they thrived in the park, and they spread throughout the region. Another 35 wolves were released in central Idaho at about the same time. Twenty years later, roughly 400 to 450 wolves inhabit the Yellowstone region, with at least 1,200 more elsewhere in the northern Rockies, and the gray wolf (that's the common name, although individuals vary in color from pale brindle to black) has been removed from the endangered species list in Montana, Wyoming, and Idaho by the U.S. Fish and Wildlife Service. Wolves are now legally hunted and trapped under state regulations in Montana and Idaho. (The situation in Wyoming, where the wolf was relisted in 2014 and hunting suspended, is more complicated.) About a hundred wolves, constituting 10 packs, live primarily within Yellowstone National Park, where Doug Smith leads the effort to monitor, manage, and protect them.

Smith has worked with wolves for 37 years, with Yellowstone's since their reintroduction, and he has handled over 500 individuals while they were tranquilized for collaring. He's a tall man with a gray handlebar mustache and crow's-feet that pinch around his smiling eyes. For all the experience, he hasn't lost his sense of gobsmacked admiration for this animal, especially when viewed at close quarters.

His colleague Dan Stahler was already there on the back side of Sepulcher, working with two other biologists on the drugged wolves. Kneeling in the snow, Stahler had almost finished fitting a collar on the bigger animal, a handsome black male, maybe three years old, with a small injury over his right eye. The other was a young female, light gray with a reddish brown head. Wearing purple medical exam gloves on a day that asked for warmer hand wear, Stahler drew blood from the male's right leg, then took a small tissue sample from the right ear for DNA work, while Smith adjusted a collar on the female. Smith measured the male: right front paw, body length, upper canine tooth—almost three centimeters (a little over an inch) for the last. What a surprise that a wolf should have very long canines. But Smith called my attention to the carnassials. "Those are shearing teeth," he said. "You don't even want to get your fingers in there when they're drugged," although that was almost precisely what he was doing, "because those just crunch bone." Carnassials are their key teeth, Smith said, edgy and powerful, for slicing meat, cracking bone. "That's what defines a carnivore, right there."

Smith and the team moved quickly. They lifted the male in a sling to weigh him: 55 kilograms, more than 120 pounds. They grabbed a fecal sample and injected a microchip between his shoulder blades. They weighed and measured the female. They took a rectal thermometer reading. Her body temperature had gone a little low, so they put her on a plastic sheet, wrapped her in jackets, and placed chemical hand warmers in her groin area while they finished the other work. When they had their data, Smith invited me to kneel in the snow beside the big male and hold up his head for a photo. Cradling the animal gingerly, I noticed that his black fur was highlighted with grizzled and silvery tips. His tongue hung out, limp as a sock. He was groggy and helpless, for now, but he was magnificent.

"Look at those eyes," Smith said. They were wide open, blazing a coppery brown. "That's wild," Smith said. "This is what our world is trying to do away with. Right here, that look. We want to keep that look. That's what Yellowstone Park is all about."

Dazed and Confused

That's what Yellowstone's grizzlies are about too. Far from tame, as Horace Albright wanted them, they are wild animals, powerful and well armed, jealous of their solitude, vehemently protective (the females) of their young. Lance Crosby's death serves as only the most recent reminder of that. They are also voracious—they've *got* to eat. Understanding the Yellowstone grizzly begins with considering its diet, and human flesh is an anomalous item, not even included with balsamroot and stink ant on that list of 266.

Kerry Gunther, the bear biologist, spoke about this one afternoon as I sat with him in the backcountry, overlooking a site that doesn't appear on the tourist maps: an odd, deep little spring that grizzlies sometimes use as a sort of bathtub. We had bushwhacked all morning to get there and ate our lunches on a small knoll, talking of what Gunther had seen in 30 years of bear study and management in Yellowstone. He's a quiet man, judicious in his statements, confident in his science, content to let others think what they will think, and dispassionate enough to view bitter disputes, with himself and other managers stuck between critics on both flanks, as "interesting."

In the 1980s, Gunther said, "every adult female bear seemed critical to the population. We were still at low population numbers." Numbers were low because the grizzly population had crashed in the 1970s, following a change in management emphasis away from Albright's yen for spectacle and toward greater attention to ecology. One signal event influencing that change was the Leopold Report of 1963, a landmark in the evolution of ideas about Yellowstone's purposes and policies, which came from a review committee chaired by Aldo Leopold's son Starker, a respected biologist in his own right. The Leopold Report, formally titled "Wildlife Management in the National Parks," wasn't the first voice to suggest an ecological approach to parks management (that idea went back to a foresighted animal ecologist named Charles C. Adams in the 1920s), but as a special advisory paper commissioned by Secretary of the Interior Stewart Udall, it carried considerable force. The report stated that conditions in each national park should be "maintained, or where necessary recreated," so as to represent "a vignette of primitive America," thereby affirming without untangling the paradox of the cultivated wild. That and the other factors, notably two grizzly-caused human fatalities (seemingly unrelated, but shockingly coincidental) in a single night in Glacier National Park, during August 1967, and the public reaction over those deaths, led to total closure of all the Yellowstone dumps.

Shutting down that garbage buffet left the bears hungry, dazed by sudden deprivation, confused, and reckless. They got into trouble, they suffered the consequences, their reproductive rate fell, and the population shrank drastically, to perhaps fewer than 140 throughout the ecosystem. During 1971 alone, 43 grizzlies were killed in various conflicts and mishaps, including 18 that a research team had captured, marked, and released. That team, led by the twin brothers Frank and John Craighead, had worked 12 years on the Yellowstone grizzly. Recognizing how firmly the bears had become habituated to garbage, and foreseeing bad consequences, the Craigheads warned against abrupt closure of the dumps (citing aspects of the Leopold Report in support—it was a complex document), but their relationship with the park ended sourly in 1971. The Yellowstone grizzly might have died out completely if the early, steep decline had continued for a decade. But four years later the grizzly bear in the lower 48 states received federal protection: It was listed as "threatened" under the Endangered Species Act of 1973. Hunting of grizzlies ceased, at least as a legal sporting activity in Montana, Wyoming, and Idaho, and the park adopted new policies to protect people from bears and vice versa.

Kerry Gunther came to Yellowstone in 1983, hired for a fishery project, and then turned to grizzlies in 1984. "We spent a lot of time managing individual bears, especially females, working really hard to try to keep them alive." That meant forestalling bear-human conflict—by practical measures such as bear-proofing garbage cans and dumpsters, patrolling campgrounds, educating visitors not to feed bears intentionally or allow them to pilfer human foods. The point was to keep humans and grizzlies at a respectful distance from each other and to encourage bear reliance on the natural foods they had begun rediscovering after closure of the dumps. It worked. More females survived, they produced more cubs, "and the population has really turned around," Gunther said. Grizzly numbers increased within the park, and their distributional range increased too, with bears now turning up in peripheral parts of the ecosystem where they hadn't been seen in decades. Grizzly bears are hard to count, but the latest estimate from the federal Interagency Grizzly Bear Study Team, within just the core area of the ecosystem (they call it the Demographic Monitoring Area)

on which they target their monitoring efforts, and using an arcane mathematical model to extrapolate, puts that population at 717 bears. Throughout the entire ecosystem, meaning Yellowstone Park and Grand Teton National Park and the surrounding wildlands, Gunther said: "I think we could easily be up around a thousand." Based on such numbers, on the trend over recent decades, and on their belief that the ecosystem is now about as full of grizzly bears as it can be, many of the state and federal bear biologists suggest that it's time to remove the Yellowstone grizzly from the list of threatened species.

This is controversial. Everything about the grizzly is controversial. Some conservationists outside agency circles have challenged the population estimates, the positive prognosis, and the advisability of delisting. They're concerned about the likely effects of renewed sport hunting, threats to bear habitat, the impacts of climate change and invasive species, and the long-term food security of the Yellowstone population. The last of those concerns, food security, extends itself into all aspects of Yellowstone and the challenges it faces.

The four natural foods on which Yellowstone's grizzlies have traditionally most depended are: cutthroat trout from Yellowstone Lake at spawning time, elk and bison meat, whitebark pine nuts, and an unusual insect that aggregates in Yellowstone's high country, the army cutworm moth *(Euxoa auxiliaris)*. Each of those is an intricate story within the big saga.

Park biologist Doug Smith races toward a tranquilized gray wolf to give it a quick physical and fit it with a radio collar before it awakens.

The Omnivore's Dilemma

The cutthroat trout is the only trout native to Yellowstone waters. Rainbow trout, browns, brookies, and lake trout are all exotics, brought in since the park was established. Yellowstone Lake was a stronghold for the Yellowstone subspecies of cutthroat, with the lake's feeder streams offering shallow, well-oxygenated waters where it could spawn. Gunther's first job at Yellowstone, as a fisheries technician, involved monitoring cutthroat spawners in Clear Creek, one of the major lake tributaries, where he remembered a record count of more than 70,000 cutthroats in a spring season, all clambering into the shallows to mix their sperm and eggs in gravelly redds. Grizzlies, black bears, coyotes, river otters, and other fish predators made good use of this easy food source. The first grizzly that Gunther ever saw was a big, dark-colored bear he encountered while snowshoeing in, at the start of May 1983, to count spawning cutthroat. Being on snowshoes was no doubt a good reminder that you should never try to run from a grizzly. By 2006 the season's count at Clear Creek had declined to 489 fish, less than a hundredth of the former number—therefore less than a hundredth the fat and protein available to bears fishing there. The decline continued, and spawning fish counts in other tributaries around the lake have fallen almost to zero.

Three factors, in deadly combination, account for this collapse: prolonged drought (especially affecting the tributaries), an ailment called whirling disease (caused by a parasite new to the system), and, most important, the presence of lake trout, brought originally from the Midwest.

Lake trout arrived in Lake Yellowstone some decades ago, secretly introduced, presumably by a witless sportsman meaning to enhance the park fishery. But the witless sportsman isn't solely responsible; lake trout had earlier been introduced to Lewis Lake and other park waters, back in the late 19th century, as a misguided act of the United States Fish Commission, a federal agency charged with protecting and promoting fisheries. That wasn't an immediate problem for Yellowstone Lake because Lewis Lake, though only six miles away, drains to the Snake River, across the Continental Divide. Still, six miles was too close for comfort. Chemical evidence from archived otoliths (ear bones) suggests that at least some lake trout from Lewis were dumped into Yellowstone Lake in the 1980s, possibly by a single numbskull with a bucket. They survived, they spawned, and their population grew, but the alarm bell didn't ring until 1994, when a fisherman on Yellowstone Lake caught a 17-inch lake trout. "Then everybody kind of had heart attacks," according to Pat Bigelow, a fisheries biologist at Yellowstone. "Because it was pretty well known that the lake trout is a voracious predator."

And a tough competitor. Big, adult lake trout in Yellowstone ate smallish cutthroat trout and fingerlings, and before long the cutthroat population was a ghost of its former self. This was not just a trade of one trout kind for another, but a drastic shift in where the trout flesh resides. Whereas cutthroats hang shallow, often eating winged insects from the water's surface, or leeches and other small invertebrates in those shallows, lake trout tend to lurk in the deep, feeding on invertebrates also, or other fish, and usually doing their spawning down there, rather than running up into the tributaries. So they are far less available to bears, and their replacement of cutthroat trout represents a severe loss for those grizzlies that once feasted at spawning streams.

Bigelow came to Yellowstone in 1979, as a young technical assistant from Vermont, and she remembers seeing fishermen on Yellowstone Lake catch and release 50 cutthroats in a day. That doesn't happen anymore. The catch rate for cutthroats is down, and if you fish there today and catch a lake trout, park regulations require that you kill it. But sportsmen will never solve this problem.

Bigelow showed me, during a long fishy morning aboard a large steel boat on Yellowstone Lake, how the park is attempting to cope with it wholesale: by contracting a company from Baileys Harbor, Wisconsin, on Lake Michigan, to bring Great Lakes commercial fishing methods to the task of exterminating lake trout. For most of five hours I stood beside her, wearing orange vinyl overalls, rubber boots, and a life jacket like everyone else, and saw the gill nets come up, bringing large greenish-gray fish, four pounds, five pounds, two feet long, that could only be considered beautiful animals if they weren't exotics causing such harm to an ecosystem. I watched her and the boat crewmen wrestle these fish from the nets, measure them, count them, then slit them open with sharp knives, check for eggs, puncture the air bladders, toss them gasping and dying into plastic tubs, and eventually dump them all back into the lake, allowing the nutrients to remain in the system. Our boat killed and dumped 238 fish by lunchtime, amounting to half a ton of lake trout. Meanwhile, three other boats were inflicting similar carnage. The work was gruesome and heartless—and justified, for the sake of not just the Yellowstone cutthroat but also the Yellowstone grizzly.

These commercial boats have been working a long season each summer since 2011, Bigelow told me, and the cutthroat have begun showing modest signs of recovery. But lake trout will never be extirpated, and so the suppression effort can probably never end. "If it fails," she said, "it'll be because we didn't try hard enough."

The loss of whitebark pine nuts from the grizzly bear menu is a more complicated concern, lying farther beyond human fixing. These fat-rich little morsels, called "nuts" but really pine seeds, come to ripeness within cones on the trees, which inhabit the Yellowstone ecosystem only at high elevations, above 8,500 feet. Whitebark pines grow slowly, reaching cone-bearing maturity after 50 years. Grizzly bears get the nuts mainly by raiding cone middens of red squirrels, which store them for winter. Clark's nutcrackers, gray-and-black birds in the crow family, also harvest the nuts and hoard them, buried in little caches of several seeds each, and most new whitebark pines grow from nutcracker caches that have gone unrecovered by the birds. That's why so many grown whitebark pines appear to have three or four major trunks, all emerging from a single base: one cache, one trunk from each seed. The greatest enemy of whitebark pine is the mountain pine beetle, a tiny bullet-shaped insect that burrows tunnels in a tree's living tissue, which can interrupt nutrient flow and kill it. Whitebark forests have always suffered episodic attacks by

mountain pine beetles, but in recent years the beetle-kill has gotten much worse, probably because of climate change. Severely cold weather, especially deep cold snaps that occur early in winter or late in spring, can knock down the beetle population; but that sort of weather is now rare, and since 2003 a vast beetle outbreak in the Yellowstone ecosystem has been killing whitebark pine like never before.

Thirty miles east of Cooke City, Montana, off the high road that climbs across Beartooth Pass in the Absarokas, lies Island Lake, with an elevation of about 9,000 feet, surrounded by boulder fields, alpine meadows, and groves of whitebark pine. Some of those trees are healthy, and some are dead, choked, desiccated, their needles gone rusty red. Standing amid them on a cool September morning, a retired U.S. Forest Service entomologist named Jesse Logan, a leading expert on whitebark and its insect enemies, briefed me on the intricacies of this situation. "The whole defensive strategy of the whitebark is escape," he said. "It's a hell of a survivor. But it's not much of a competitor." At high elevations, in cold and harsh conditions, whitebark largely escapes competition from other conifers—ponderosa pine, lodgepole, Douglas fir. But it doesn't escape the beetle, not anymore.

Logan spent much of his career on pines and their insect enemies. His aerial surveys, done with a colleague, William MacFarlane, from Utah State University, suggest that almost half of the whitebark distribution in the Yellowstone ecosystem has now suffered severe

Jackson Lake is the biggest body of water in Grand Teton National Park, made more magnificent because it rests at the foot of the Tetons.

mortality from mountain pine beetle, and 82 percent has suffered at least moderate die-offs. He pulled some bark off a dead tree to show me the beetles' tunnels, vertical and crisscrossing, like a subway map of New York. As the warming climate rises upslope to these high elevations, allowing beetle outbreaks that never end, where else can whitebark go? "My sense is that the loss of this food resource is really important to grizzlies," he said. He's not a bear biologist, and others disagree, but the issue is serious.

Army cutworm moths are more unexpected, more fortuitous in the grizzly bear's diet because, unlike whitebark nuts and cutthroat trout, they come from elsewhere. These little gray moths, familiar annoyances around porch lights and screen doors on July evenings in Cody, Wyoming, and commonly called "millers," migrate hundreds of miles in early summer from lowland farming areas—on the Great Plains and in the Intermountain West, where in the larval stage they are crop pests—to high elevations in the Absarokas. They spend the hot days hunkering in cool, moist recesses amid scree slopes (fields of broken, tumbled rocks) above 9,000 feet, and at night they fly out to drink nectar from wildflowers on alpine meadows. Metabolizing the nectar, they lay on rich stores of fat, enough to see them through their arduous return migrations, in autumn, back to crop fields in Kansas, Nebraska, or wherever. A moth that arrives in the mountains in June, having 40 percent body fat, can increase to 65 percent fat by September. For a grizzly bear, lapping up such creatures from amid the scree is like eating pill capsules filled with olive oil by the handful. A grizzly can consume up to 40,000 moths in a day, representing about 17,600 calories, which makes the moths ideal food for the hyperphagia period of the bear's year, when it's fattening itself for winter hibernation. At that rate, a grizzly feeding for 30 days on army cutworm moths can satisfy almost half its yearly energy needs.

One of those moth sites lies on the northwest face of a mountain, high above a beautiful little basin, near the headwaters of the North Fork of the Shoshone River. On a cool day in late August, Mark Bruscino, formerly chief of large carnivore management for the Wyoming Game and Fish Department, led a small group of us to the site, seven hours by horse into the North Absaroka Wilderness. Our pack train had been assembled by Lee Livingston, an outfitter and county commissioner out of Cody, who rode caboose. We made camp in a glade, drank some whiskey, ate well, and awoke the next morning to find that a small, late-summer storm had covered the higher slopes with a thin blanket of snow.

We stood at spotting scopes all morning, watching seven grizzlies work their way across the whiteness of a scree slope far above. The bears were burly and dark, their paths turning scree-brown against the whiteness as each bear tossed aside 40-pound rocks, digging down, slurping moths, moving slowly on. For a while I watched one grizzly, his huge butt protruding in the air as he dug, rummaged, snarfed, and then occasionally lifted his head, as though to take a breath and swallow. Not far from him, a female and two cubs also fed, tolerating the male at what would be unacceptably close proximity (because male grizzlies kill cubs and eat them) under other circumstances. But they were all preoccupied now, gobbling moths, acquiring fat.

So far this year, Bruscino told me, his old colleagues had counted about 200 individual grizzlies working dozens of different moth sites. That's about a fifth or a quarter of the grizzly population throughout the entire Yellowstone ecosystem, he reckoned, all feeding on moths while they're available. His arithmetic assumed, as did Gunther's, about a thousand grizzly bears—a huge rise since the nadir of the 1970s. "To me," he said, "this is one of the greatest wildlife restoration stories ever."

The grizzly still faces threats, Bruscino noted, but the worst is not the loss of whitebark pines nor of cutthroat trout. It's not that the moths might disappear, or become poisonous, if farmers in Kansas decide to use more pesticides. "The biggest long-term threat to this bear population is private lands development," he said.

The Last-Son-of-a-Bitch Clause

Private lands—especially in the form of big cattle ranches, lying just off the plateau—are crucial not just to grizzly bears but also as winter range for many of Yellowstone's elk. That those elk migrate seasonally, from green highlands in summer to lower elevations for the cold months, has long been known. But the intricate pulsings of different Yellowstone elk herds are newly illuminated by several researchers, including a young ecologist named Arthur Middleton. Middleton was hard at work, in the Thorofare area outside the park's southeastern boundary, when I rode over the mountains to find him.

My guide this time was Wes Livingston, Lee's brother, a crusty backcountry horseman with uncommon savvy about how to navigate this wilderness and cope with trouble. Wes wore a camo T-shirt, a droopy handlebar mustache, an elk-tooth necklace, a weathered felt hat that didn't even bother to look cowboy, and a titanium .44 magnum on his belt, in case he had need to shoot an injured horse or mule. He led a string of five pack mules and an extra horse as I followed him up a long switchback trail toward Deer Creek Pass.

It was late July but, nearing the top, we faced a dicey stretch where the narrow trail still lay buried beneath a drift of old snow. Off the left side was a steep plunge toward Deer Creek, far below. Wes stampeded his animals across, their hooves post-holing into the snow, their momentum carrying them through. As I tried to follow, my horse—a steady buckskin named Jimbo—balked, floundered, and got himself turned. "Get off!" Wes hollered, then coached me to grab Jimbo's lead rope and walk him back onto the dirt to regain his footing. (This was the same spot where, a week later, coming out, we would see a dead horse 60 feet below the trail, fallen from another pack string and half-eaten by some grizzly.) I followed instructions, and eventually we all got across.

At the top of the pass we noticed a scoop shovel, stashed there for the use of snowbound hunters. Then the trail tipped down gently into a wide, meadowy valley on the Thorofare side, and Wes laconically declared—out loud, but to himself—his satisfaction: "Didn't roll any donkeys down the mountain." I wasn't sure whether "any donkeys" referred to the mules, the horses, or me.

We joined Arthur Middleton at his camp near the mouth of Open Creek. The following morning, over a campfire breakfast, he and Wes began discussing elk population dynamics and migrations, a conversation that would continue throughout our trip. The northern herd in Yellowstone Park is famous, Middleton said, for its fluctuations between excessive abundance and relative scarcity, due partly to natural factors, including now again wolf predation, and increased bear predation, and partly to direct human actions. That's the herd most familiar to visitors, thanks to its conspicuousness along roadsides, from Hayden Valley to the grassy lawns around Yellowstone's headquarters buildings at Mammoth. The other herds probably haven't experienced such drastic ups and downs, for reasons that might include open terrain, long sight lines against predators, strong winds to blow away snow, and the animosity of private landowners toward wolves.

Attitudes regarding the wolf are more bitterly polarized than those around any other issue in Yellowstone. Even beyond the wolf-haters-versus-wolf-lovers tussle, serious disagreement exists among scientists about how, and to what degree, wolves are reshaping the Yellowstone ecosystem. Do they reduce reproductive success among elk simply by creating a "landscape of fear," wherein the great bulls and cows are too nervous to eat and procreate? Have wolves killed enough elk to curtail elk browsing on aspen and willow shoots? Has that reduced browsing allowed aspen and willow stands in Yellowstone to recover and renew themselves for the first time in decades? Has such aspen and willow recovery enabled the return of beavers and songbirds? Or is reality a little more intricate? Some scientists and wolf advocates tell this story in happy, simplistic terms. "But it's an unproven theory that gets undue attention," Middleton said, "in the quest to have wolves shine rainbows out of their asses."

Arthur Middleton is an improbable fit for the role of Wyoming elk maven: a kid from a fancy eastern prep school who did a master's degree at the Yale School of Forestry and Environmental Studies, then came west in 2005, having landed work on a study of elk-wolf interactions commissioned by the Wyoming Game and Fish Department. Arriving in Cody to meet his new collaborator, he admitted that he'd never *seen* a wolf or an elk. But he learned fast, and he loved the mountains. He put global positioning system (GPS) collars on elk, clarifying poorly understood patterns of their movement between summer and winter ranges. He collated similar data from other researchers and made eye-opening digital maps. Look where these animals *go*. "Most of Yellowstone's elk," he said later, "are not *in* Yellowstone for most of the year." They are off the plateau, down on winter range, where the snow isn't so deep and the temperatures aren't so brutal, largely on private ranches. Middleton's elk work began as a Ph.D. project through the University of Wyoming and would continue on a postdoc fellowship from Yale. His hair was long, his speech was slow and considered, and his brow scrinched when he pondered something carefully. Within a short time, at least one Cody-based game warden started referring to him as "the elk hippie."

Middleton's mapping of the elk data showed at least nine distinct migratory herds, each moving seasonally in a different direction.

Many of Grand Teton's elk winter on the National Elk Refuge, just north of Jackson, Wyoming. Thousands from the other herds move down onto big private ranches. The massive slaughter in the late 19th century took numbers way down, followed by the elk boom under Horace Albright's protection and then a policy reversal after elk seemed *too* abundant, especially on the northern range of Yellowstone Park, resulting in an active elk-reduction program that lasted from 1934 to 1967. Whipsaw changes. During that long elk-reduction regime, park rangers shot 13,753 elk from the northern herd, private hunters killed 41,400 when those animals migrated out of the park, and almost 7,000 were trapped and shipped away to forests and zoos elsewhere. Then came the late 1960s, when the park superintendent and his chief biologist, influenced by the Leopold Report and some fashionable new thinking in ecology, embraced a policy called "natural regulation." That served to codify in two words—but not solve—the paradox of the cultivated wild. Just what the words "natural regulation" might mean in a given instance was unclear, and the policy has never been an absolute one.

As for elk, natural factors such as drought and fire, and indirect human influences such as climate change exacerbating drought and fire, plus the wolf reintroduction, plus the planting of lake trout in Yellowstone Lake (with consequences for cutthroat trout and grizzlies),

---

Downstream from the Oxbow Bend of the Snake River, Schwabacher's Landing is known for its exceptional wildlife watching and sightlines to the zeniths of the Tetons.

have affected their numbers and migration trends too. Those are some of the ramifying effects that Middleton wants to understand.

Several days into the Thorofare trip, we rode up out of the valley and camped high, at about 9,200 feet, atop the Thorofare Plateau. Next morning Arthur and I went higher, picking our way through a jackstraw mess of downed timber from the 1988 fires, our horses stepping carefully over the logs, then onto the easier footing of a meadow, amid the short grasses, the Indian paintbrush, the purple asters. Cresting a ridge, we saw about 150 elk—bulls and cows, some calves—grazing and resting on a green slope in the near distance. Part of the Cody herd, Arthur said. Between their winter range and this good summer grazing, they had traveled about 50 miles and crossed two steep mountain passes. Part of what he wanted to do here, Arthur explained, was gauge the cow-calf ratio. If 80 percent of the cows had calves in spring, and only 40 percent do now, where have those other calves gone? How many have been eaten by grizzlies, how many by wolves?

His other keenest interest is their migration needs. "Imagine if the boundary of Yellowstone National Park was a wall," he said to me later. Raise the wall in April, and you would "basically exclude these animals from the park." They couldn't come up from their winter range, as those lands became hot and dry, to eat the nutritious young summer grasses atop this plateau. Raise the wall in October, trapping the elk *inside* Yellowstone, and you'd see "fairly dramatic die-offs during the winter months," as they starved and froze within their high-elevation cage. Also, there would be secondary consequences. Yellowstone's vegetation would suffer from overgrazing, even more so than it already does.

Carnivores and scavengers would benefit in the short term, with so many elk dying, and then suffer their absence in the long term. Tourists would miss seeing the elk and, especially in the Lamar Valley, the wolves that stalk them. Grizzly bears would need to replace another lost food. Hunting opportunities would decline, outfitters like Lee Livingston would suffer, and (because elk hunting is a major revenue engine) businesses would close on Sheridan Avenue in Cody.

Of course, no one is going to raise a wall around Yellowstone National Park—except perhaps incrementally, by way of private lands development that turns winter range into vacation homes for the wealthy, sprawling suburbs for the middle class, and commercial areas that serve all sectors of the region's growing human population. That's generally what happens when the big, old private ranches, the historic cattle operations, change hands in the modern era—as Mark Bruscino had warned. Many people understand this, at least dimly, but want to own a piece of the scenery anyway, to erosive effect. Wes Livingston made that point pungently, as we sat drinking coffee: "It's called the Last-Son-of-a-Bitch Clause: 'Let me in, and then close the gate.'" Blocking the migration corridors, by settlement on lands just outside Yellowstone and Grand Teton, will interrupt an essential flow of ungulates and all the values (nutritional, ecological, aesthetic, financial) they carry. It's the same basic truth that Phil Sheridan recognized back in 1882, further illuminated by research such as Arthur Middleton's: Yellowstone's elk need the full ecosystem, including winter range outside the park, and the ecosystem needs Yellowstone's elk.

Wes had a piece of advice, he told me, for the righteous, out-of-state greenies he sometimes met, affluent people with their vacation home sites and little hobby ranches carved out of these mountains. If you really want to help Yellowstone wildlife, he'd say, "Burn your house down and go back to California."

These Places Are Huge

Among those big private spreads around the edges of the Yellowstone ecosystem, so crucial collectively to Yellowstone's elk and other wildlife, is the TE Ranch, once owned by Buffalo Bill Cody himself. The TE sits wide and quiet across the verdant upper valley of the South Fork of the Shoshone River, southwest of the town of Cody and beyond the sprawl of its suburbs. Buffalo Bill acquired the core of the place in 1895, at the height of his success as a Wild West showman; he also bought some horses branded TE, as the story goes—"TE" for Trails End or who knows what—and adopted that brand rather than rebranding the horses. He added thousands more acres but then lost a good part of his holdings in financial troubles near the end of his life. The history of large ranches in the northern Rockies reiterates that pattern—a smallish place acquired by homesteading or purchase, followed by seized opportunities as neighbors gave up, further land purchases and consolidation, followed often by later reversals or changes of circumstance leading to parceling the place away among children or sale for subdivision or other forms of shrinkage. Only the luckiest, most obdurate, or steadiest families managed to assemble a large ranch and keep it large for succeeding generations.

Bill Cody's ranch passed through several owners before its purchase in 1972 by Charles Duncan, Jr., a Texan then president of Coca-Cola. In 1975, Mr. Duncan bought a second ranch, adding about 10,000 acres, and later acquired still another. Also, like most western ranchers, he had grazing leases on Forest Service and Bureau of Land Management lands. In 1988, he hired a young Cody man named Curt Bales, just 27 years old, to manage the operation, and Bales still holds that position today. As a lean middle-aged man in a cowboy hat and a western shirt, with a computer on his desk and a view of the mountains beyond, Curt Bales welcomed me for a chat at the ranch office.

Yes, he grew up on a family ranch himself, Bales told me, a smaller place closer to town, now operated by his two brothers. Ranching, it's gotten harder since then. Beef prices have gone high but costs even higher. The Bales place once fed five families; nowadays it feeds two. So a young person, a third son as Curt was, who wants to stay in ranching somehow, finds a job—and if he's very lucky, earnest and dependable, a job such as his at the TE. "Right now we're running about seven hundred head of mother cows." Grazing conservatively, he said. But of course that number doesn't count the elk, with their own toll on the grasses, their own costs to the operation. According to one study, done in the mid-1990s, the TE Ranch in March and April hosted 1,700 head of public elk. And the elk attract wolves. The wolves follow the elk down

to winter range, but then the wolves den, and when the elk migrate back to the highlands in spring, the wolves with their dependent pups remain where they have denned, feeding sometimes on cattle. Only with state compensation for such losses, plus patience, can ranches absorb these costs.

But there are felicities too, if you're a Cody boy who loves the region, to living and working on this wildish landscape, he acknowledged. "We have everything here on the ranch that Yellowstone Park has, with exception of maybe the buffalo. I mean, we've got grizzly bears, we've got the wolves, we've got mountain lions, we've got bighorn sheep, we've got the elk, we've got the deer, we've got the moose." And no tour busses unloading crowds of people, all eager to take pictures of themselves in front of a famous geyser.

When he first came here, Bales said, he had his reservations about absentee owners and the big ranches just getting bigger. "But I really changed my tune." He began to understand that those owners have a love for the landscape too, and that if it wasn't for them and people like them today, "this whole valley would look like the North Fork." The North Fork of the Shoshone in its lower reaches, the valley through which Highway 14 leads from Cody to Yellowstone Park, is a promenade of motels and tourist lodges and restaurants and homes on small plots pushing out as far as private lands will allow, right up against the national forest. The upper South Fork road, by contrast, leads nowhere but to a dead end along the river, passing the TE and just a few other large private ranches on its way.

"These places are huge for the existence of wildlife," he said.

The Creeping Crisis

D ave Hallac stood in his office at the Yellowstone Center for Resources (YCR), the park body charged with science and resource management, which is housed in a rambling old clapboard building amid the formidable stone structures of Mammoth. A smart, candid man with an oval face and thinning hair who was originally from New Jersey, Hallac was serving his last day as chief of the YCR—head scientist at Yellowstone Park— before departing to a promotion elsewhere. His office had the look of farewell when I found him there, between final tasks, his shelves and desk already bare, his books and reports and photographs packed into boxes awaiting removal. He closed the door, which was a little unusual on that open-door corridor, and as we sat amid the boxes, he repeated to me something he had said passingly a couple of months earlier— something so arrestingly blunt that I had asked him to elaborate. "I think we're losing this place," he said. "Slowly. Incrementally. In a cumulative fashion." He hesitated. "I call it a sort of 'creeping crisis.'"

Hallac ticked through a list of interrelated concerns, nagging issues in Yellowstone and familiar to us both: bison management, elk migration, grizzly bear conservation, private lands development in the region surrounding the park, human population growth driving that development, invasive species and their impacts on native species,

Playing with one's food may not be polite for people, but this coyote cannot help it after finding a tasty vole in a meadow full of exotic dandelions in the Lamar Valley.

BOB COCHRAN

water use, climate change, and finally the overarching problem that exacerbates all these others: an absence of coordinated, transboundary management among the various agencies, civil entities, and owners controlling their pieces of what has come to be recognized as the Greater Yellowstone Ecosystem. Fragmentation of responsibility, divergent goals, crosscutting efforts, dissipation of vigilance. "We go around telling everybody this is the most intact ecosystem in the lower 48," Hallac said. "Well, if it's that important, that special, it's time for us to do a lot better when it comes to protecting it."

This worry about the broader wholeness of Yellowstone, which others share (though not enough to leverage vigorous action), is now urgent but has been a long time coming. The word "ecosystem" didn't appear in the 1872 act establishing Yellowstone National Park, and probably not in any of the emendations or directives about the park that followed for much of a century. Ecological thinking did come to Yellowstone earlier, back in the 1920s, thanks to the ecologist Charles C. Adams among others, but it was then a marginal perspective overshadowed by emphasis on tourism development and protection of the "good" animals, such as elk and songbirds, versus the "bad" animals that preyed on the good, such as mountain lions, wolves, coyotes, and the white pelicans that nested on Molly Island.

The phrase "greater Yellowstone ecosystem" may have been first used in the 1979 book *Track of the Grizzly,* by Frank Craighead, an account of the path-breaking 12-year field study of bears led by him and his brother John. The Craigheads had absorbed, and stated pointedly, a crucial fact: that Yellowstone's grizzlies live not within the park's boundaries (which are unfenced and, throughout most of the landscape, unmarked) but across a wider terrain that includes also Grand Teton National Park, parts of adjacent national forests, and other surrounding lands. Two years later, the superintendent of Yellowstone, a percipient man named John Townsley, used that phrase during a friendly chat with a young mountaineer and educator named Rick Reese. "He told me that to treat Yellowstone Park as a box on a map, with no regard for threats to the park from neighboring national forest lands, was absurd," Reese later wrote. The park and its animal populations, its plants, its waters, even its thermal features, would be affected by what happened outside that box. Paraphrasing Townsley, Reese wrote that "The American people must be educated about these interrelationships and must begin to think in terms of a 'Greater Yellowstone Ecosystem.'" John Townsley died a year later, but Reese and others took it from there.

Reese himself served as first president of a new organization, the Greater Yellowstone Coalition, an alliance of individuals and groups dedicated to conserving the broader wholeness of Yellowstone. Soon afterward, in 1984, he published a book, graced with pretty photographs but made influential by serious words, titled *Greater Yellowstone: The National Park and Adjacent Wildlands.* For a while some of the government people resisted this new term, favoring instead "the Greater Yellowstone Area," a pusillanimous compromise, possibly because "ecosystem" invoked an interconnectedness that ran against bureaucratic compartmentalization and discomfited those with a jealous sense of turf. But by now even they, except the most cautious, have adopted it.

The Greater Yellowstone Ecosystem (GYE) is an amoeboid expanse of landscape encompassing Yellowstone National Park, Grand Teton National Park, the John D. Rockefeller Jr. Memorial Parkway (a connector strip between the two parks, also administered by the National Park Service), the National Elk Refuge in Jackson Hole, Red Rock Lakes National Wildlife Refuge, a portion of the Wind River Indian Reservation, portions of five national forests and some Bureau of Land Management holdings, as well as private lands under the political jurisdictions of Wyoming, Montana, and Idaho. The whole shebang amounts to about 22.6 million acres, within which prevails (imperfectly in some areas, forcefully in others) an ambiance of Rocky Mountain wildness. Surrounding this great amoeba is a modest transition zone, where you will more likely find cattle than elk, more likely see a grain elevator than a grizzly bear, and more likely hear the bark of a black Labrador than the howl of a wolf. Surrounding that buffer is 21st-century America: highways, towns, parking lots, malls, endlessly sprawling suburbs, golf courses, Starbucks.

Who's in charge of the Greater Yellowstone Ecosystem? Everybody and nobody. There is a deliberative body, the Greater Yellowstone Coordinating Committee, but that organization includes only representatives of the federal agencies, not the states, county officials, or private interests, and its powers are modest even when its consensus is firm. That's why Dave Hallac bemoaned the absence of transboundary management in the face of the creeping crisis. "I'm not suggesting in any way that the ecosystem is falling apart," Hallac told me. "Because it's not. But it has the potential to. And there are a

A mother badger emerges from a burrow near Mormon Row in Grand Teton National Park. Next to wolverines, American badgers are the largest members of the weasel family.

RONAN DONOVAN

lot of incremental changes that, I think, are beginning to negatively affect the system."

After half an hour of such conversation, someone knocked on Hallac's door and peeked in, reminding him that he was due at his own farewell party. He invited me along, but I didn't want to intrude, preferring to skulk away and consider what he'd said. Park managers come and go, even the best of them, but the problems abide. ∎

On the lookout for
raptor predators,
a yellow-bellied mar-
mot emerges from the
safety of its den and
scouts its rugged sur-
roundings for food.

BARRETT HEDGES

(opposite)
**Even before winter
has fully receded,
mating pairs of sand-
hill cranes arrive in
Yellowstone to choose
nesting sites. Their
mating calls are
unmistakable, their
devotion to one
another inspiring.**

MICHAEL FORSBERG

*Blending into the tall
grass, a great gray
owl lies in wait for
dinner. Rodents, even
young hares or rab-
bits, are favorite sta-
ples of the great gray.*

RONAN DONOVAN

While a human presence proliferates around the world, parts of the Yellowstone region are wilder than they've been in a century. Grizzlies are spreading their range, for example. Here, a bear in Grand Teton National Park fends off ravens from a bison carcass dumped by rangers—a scene that's both wild and not.

CHARLIE HAMILTON JAMES

Detail

The Complex Food Web

In the drama of eat and be eaten that connects all living organisms, keystone species such as bears, elk, and cutthroat trout play a crucial role.

For decades, particularly before the region sported restaurants, visitors enjoyed fishing for their dinners. But supply was limited: Nearly half the park's lakes and streams lacked fish. To protect the scarce stock, park managers killed fish-eating wildlife, such as bears and pelicans. They also asked the U.S. Fish Commission to stock the waters "so that the pleasure seeker can enjoy fine fishing within a few rods of any hotel or camp."

Between 1881 and 1955, 310 million fish were planted in Yellowstone, wreaking havoc on the ecosystem's equilibrium. Native fish—such as the Yellowstone cutthroat trout, a source of food for dozens of animals—were displaced, attacked, and bastardized by non-native species. From 1978 to 2010 cutthroat numbers in the lake declined even more precipitously, owing to a combination of drought, parasitic disease, and competition from invasive species. As cutthroat numbers fell, predators found other food, impacting those species. Grizzlies took more elk. Bald eagles' prey came to include swan cygnets.

Today the priority is on protecting the park's 13 native fish types. Strict catch-and-release rules governing rod-toting human and management practices are helping the native fish rebound.

With winter advancing fast, a Yellowstone grizzly bear, caught via a camera trap, breaks into a cache of whitebark pinecones stashed by red squirrels.

DREW RUSH

Following a blizzard in the Custer-Gallatin National Forest near Bozeman, Montana, a pine forest is covered in marvelous crystalline snow sculptures.

DAVID GUTTENFELDER

Once nearly wiped out by fur trappers, river otters have rebounded and now are common sights along rivers such as the Snake in Jackson Hole. Here, an adult otter and companion feast on cutthroat trout.

CHARLIE HAMILTON JAMES

(opposite)
Magpies, members of the crow family, are especially adept at shadowing predators. Here, a bird tries to steal fish from a clan of otters. If the magpie isn't careful, it could become lunch.

CHARLIE HAMILTON JAMES

The carcass of a bison that drowned in the Yellowstone River became a feast for the alpha female of the Mollie wolf pack and her two-year-old offspring. Bringing down a live one-ton bison is dangerous; Yellowstone wolves far more often target elk, which make up 85 percent of their diet.

RONAN DONOVAN

The Greater Yellowstone Ecosystem is globally famous for its native cutthroat trout, highly prized by anglers, but climate change threatens to severely impact these cold-water relics from the Ice Age. Here, cutthroats spawn in a tributary to the Gros Ventre River on the edge of Jackson Hole.

CHARLIE HAMILTON JAMES

(opposite)
The illegal introduction of lake trout has devastated Yellowstone's cutthroat trout population. Targeted for aggressive control, here a lake trout has been caught in a gill net.

CHARLIE HAMILTON JAMES

Invasive lake trout can grow huge, as fisherman Trevor Beutel demonstrates (above), fattened up by eating the native cutthroat, as found in the belly of a netted lake trout (below).

CHARLIE HAMILTON JAMES

(following pages)
A camera trap set along a game trail in Grand Teton National Park glimpses many creatures: a mule deer fawn (left), a black bear, and a mountain lion (right), known to have a vast territory but here sharing the pathway with others.

CHARLIE HAMILTON JAMES

John Craighead

WILDLIFE BIOLOGIST, PIONEER OF GRIZZLY MANAGEMENT

"My God, I'd never seen anything like it. Frank and I had seen plenty of beautiful mountains in Pennsylvania, but the day our '28 Chevy topped this hill in Wyoming and we spotted the Tetons, it was like our souls got sucked right into the Rocky Mountains. We knew right then and there that our calling was out West, and that any professional endeavor would have to somehow be centered in those mountains."

from Grizzlies and Grizzled Old Men *by Mike Lapinski*

John Craighead and his late twin brother, Frank, were pioneers of grizzly bear research in the Greater Yellowstone Ecosystem, advancing the use of special collars for tracking the bruins via radio telemetry. The Craigheads wrote several stories on bears and raptors for National Geographic *magazine.*

ERIKA LARSEN

PART THREE

Beyond the Frame

If the national park idea is . . . the best idea America ever had, wilderness preservation is the highest refinement of that idea.

—WALLACE STEGNER, 1990

Crossing the Thorofare Plateau, the Cody herd—one of nine major wapiti, or elk, subpopulations in Greater Yellowstone—moves from the high country toward areas at lower elevation for the winter. Here, big game animals can still migrate long distances, a valuable characteristic of this magnificent ecosystem.

JOE RIIS

Abducted by Aliens

Some problems can be fixed. Some mistakes can be rectified. The extirpation of wolves from Yellowstone National Park and the vacancy of

the wolf's ecological niche fall in that category. Seldom in United States history has such a bold conservation initiative yielded such a resounding and controversial success.

The first wolves that recolonized Yellowstone, after 60 years of wolf absence, were captured in western Canada during January 1995, by tranquilizer darting from helicopters, and transported south in large wooden shipping crates by plane and then by truck to the Lamar

No longer lone, an adopted member of the Phantom Springs wolf pack stands tall in Grand Teton National Park. After an absence of about 70 years, wolves returned to the park in 1998, moving down from Yellowstone.

Valley. It must have been a harrowing journey. For 10 weeks they lived in acclimation pens, largish areas behind high fencing, each pen enclosing roughly an acre. There were three pens, all located discreetly up little drainages beyond sight from the Lamar road—one at Rose Creek, one at Crystal Creek, one farther upstream at Soda Butte—and containing a total of 14 translocated wolves. At first the wolves scarcely dared step out of the crates. They had been abducted by aliens, after all, and who knew what might happen next? Packmates from Canada, already familiar to one another, were assigned to the same pens, in an effort to minimize trauma. They ate roadkill deer, elk, moose, and bison brought on mule-drawn sleds by park biologists, who otherwise left them alone.

After their 10 weeks' acclimation, the fences were opened and the wolves, tentatively at first, wandered free.

Not all of them survived the hazards of their new location (one alpha male, known as #10, ranged up into Montana and was promptly, illegally, shot) but most did. Abducted or not, they made the best of the situation. During the second year, 1996, another 17 wolves arrived and two new acclimation pens were opened in other parts of the park. That spring, those animals too walked free, and wolf reintroduction was essentially complete.

The chief of this Yellowstone Gray Wolf Restoration Project was Michael K. Phillips, an experienced carnivore biologist who had worked on red wolf recovery in North Carolina. Doug Smith was then a field biologist following Phillips's lead (before succeeding him, in 1997, as head of what's now simply the Yellowstone Wolf Project). Phillips, a believer in well-calibrated public outreach, offered a few journalists the opportunity to tag along and glimpse the acclimation process. I was among the lucky, and so on a March morning in 1995, before the first releases had occurred, I post-holed uphill through the snow, alongside Phillips and his mules and his sled-load of frozen elk carcasses, to the Rose Creek pen. I remember helping unload the meat, dragging it into the enclosure, and then being invited by Phillips to peer into one of the transport crates, which had been left as a sort of doghouse. An adult female wolf, huddling inside, peered back like a worried puppy.

Almost 20 years later, not many miles from that spot, I stood with a small group of hardy souls, devoted wolf-watchers in wool hats and gloves, equipped with spotting scopes on tripods, as we gazed out across the valley toward a large, gray wolf on an elk carcass near Soda Butte Creek. Ravens attended nearby. It was just after dawn, early May, and cold. These people were cheerily seizing the day. A unique culture of such amiably fanatical wolf-watchers has arisen since 1995, and the Lamar Valley is their headquarters; you can find some of them any day of the year, clustered at turnouts along the Lamar road. They are generous with their pooled information and with views through their scopes—to a newcomer like me, for instance, arriving with only binoculars. This gray adult on the elk, he's the alpha male from what's now called the Lamar Canyon Pack, someone explained. The female is in her den, with the cubs, up on that hillside to the north, in the trees. Moments later we were joined by Rick McIntyre, in his park service olive, the chief observer and indefatigable chronicler of the wolves of the Lamar.

McIntyre, a soft-spoken man with sandy red hair sticking out from his cap and extraordinarily pale skin despite living his life outdoors, has spent much of the past two decades watching and recording the behavior of these animals—their pairings, their tiffs, their acts of predation and parenting, their conflicts between one pack and another, their transfers of allegiance, their competitions with grizzlies and mountain lions, their thefts of carcasses, their play, the growth and maturation of pups, the aging of adults, the sad but inevitable disappearance of those who can no longer keep up. Back in 1994, McIntyre was a naturalist ranger who did interpretive programs, carrying a wolf pelt into busy areas such as Old Faithful and Mammoth, drawing a crowd, and talking about wolves. Then the live wolves arrived. For almost two decades now he has worked for Doug Smith on the Yellowstone Wolf Project, though his job is still interpretation as much as monitoring. He's an ambassador for the wolves, adding guidance and understanding to the way Yellowstone visitors see them. He knows the animals, individual by individual, pack by pack, their genealogies and their proclivities, as thoroughly as if he had written the script of this great canine soap opera himself. The female in her den at Soda Butte, he told me, is the fifth-generation descendant of the Canadian pair who acclimatized in the Rose Creek pen. With that statement, I realized: On that day with Mike Phillips 20 years ago, I probably looked her great-great-grandmother in the eyes.

No one foresaw, McIntyre said, that wolves would settle in so comfortably and be so easy to watch in the Lamar Valley. But in retrospect it's explicable: The habitat is excellent along both sides of the little road, the prey is plentiful, the sight lines are long, and the animals here—unharried by hunters—don't spook at the sound of a car or the presence of humans with tripods and scopes. They go about their business, sniffing and stalking, killing and eating, mating and raising their young, and occasionally howling, while people from all over the world stand rapt at the roadside, enjoying this rare chance to see wolves in the wild. According to a 2006 study by economist John W. Duffield and two colleagues at the University of Montana, wolf-focused tourism by visitors to Yellowstone at that time brought an estimated $35.5 million annually to the economies of Wyoming, Montana, and Idaho. Park visitor numbers have since increased substantially, and the cash value of live wolves presumably has too. The Duffield team's survey data placed wolves second on the list of animals that people most wanted to see, behind only grizzly bears.

Wolves had meanwhile been reintroduced also to a wilderness area in central Idaho, almost 200 miles west of Yellowstone. From both reintroduction areas they spread, multiplied, and spread farther, reoccupying wolf habitat to the point that, within two decades, at least 1,600 wolves composing some 280 packs lived in the three-state area. Occasionally they preyed on livestock as well as wild ungulates, getting themselves in trouble, provoking control measures (wolf killing) by the U.S. Fish and Wildlife Service, and reawakening among ranching communities a vehement wolf-hatred that had lain dormant for decades, like the deep, sore memory of a blood feud. For reasons that even economics and psychology can't untangle, some people truly hate wolves, in a way no one hates grizzly bears or mountain lions. Nevertheless, wolves were back and thriving, with their passionate defenders (such as the watchers in the Lamar) as well as their passionate despisers. Lawsuits, court decisions, and proposed rules for removing them from listing under the Endangered Species Act flew every which way. Then, in 2011, the wolves in Montana and Idaho were delisted (while the Wyoming situation remained in dispute), whereupon those two states began to license hunting and trapping.

Wolfers

Just down the valley from Soda Butte, Rick McIntyre and I met a group of visitors that included Elli Radinger, a wolf devotee, author of nature books, and tour operator from Germany. Radinger began leading wolf-focused tours to Yellowstone back in 2000, she told me, and continued at the rate of about three or four per year until recently. Since delisting occurred, though, and her German clientele learned that the Lamar wolves could now be legally shot when they wander north across the park boundary into Montana, that portion of her business has crashed. "People have a very personal relationship to these Yellowstone wolves," she said. "They found out about the hunting outside—and especially about the killing of some of our favorite wolves. And they said, 'We're not going there anymore.'" It is, she added, "really a very emotional thing."

Just beyond the park boundary lies remote terrain that, according to Montana's administrative map for hunting and fishing, falls within Region 3. One of the state's wolf biologists for Region 3 is a young woman named Abby Nelson, who grew up riding horses in Greenwich, Connecticut, before coming west to Colorado College. Nelson studied veterinary science and wildlife biology, getting a master's degree from the University of Wyoming and spending three winters as a volunteer, then further time as a hired tech on Doug Smith's wolf program in Yellowstone. She began work for the state of Montana in 2010, trapping wolves to affix radio collars, following their movements throughout her region, monitoring their fates.

When I visited her camp on West Rosebud Creek, one of the most hidden-away corners of Montana, she was trapping for the Rosebud Pack—a mating pair with three new pups—in order to get a collar into that group. She had six leghold traps set on her trapline, each with padded jaws, carefully buried, and tactically located to intercept the wolves where they walked. I rode along on the back of her four-wheeler while she checked the traps, all empty that morning. Then she placed a seventh trap, setting the jaws, laying it into a recess she had dug, covering it with a screen, then with finely sieved dirt and pine needles, making fastidious adjustments with a whisk broom, adding canine urine (a territorial attractant) from a bottle, all while standing on a small canvas tarp and wearing gloves, to avoid leaving any suspiciously human scent. They don't teach these skills in Greenwich.

The wolf population in Montana had risen above 600, she told me, despite fairly liberal hunting and trapping quotas, with 230 animals taken in the past year. The inherent growth rate was such that only sizable harvest would keep the population size within the state's desired limits. Part of her job was to verify and record each wolf death within her region. "It's hard to kill wolves," she said. "And the thing about wolves, whether you like them or hate them—they're so fecund, they'll still hold their numbers." The average wolf litter is about five and a half pups, and an alpha pair can produce a litter each year.

A wolf pack follows grizzly bear tracks through the snow to what might be the next meal, likely to be elk.

RONAN DONOVAN

Months later, over coffee in Bozeman, I asked Nelson whether she felt any conflict between the biological and the emotional dimensions of her job. "I'm a trained biologist," she said. "We're trained to be objective, and to manage wildlife populations rather than individuals." Wildlife management, she added, necessarily involves "the human social ecosystem" as well as the wildlife itself. "Wolves in particular are so polarizing." She groped for words, then found some: "You can't make progress on anything until, you know, whatever issue it is, it becomes a little less polarized. And I think that takes compromise."

Abby Nelson followed up on our conversations, months later still, by introducing me to one of the wolf killers whose licensed activities fall within her regulatory ambit. This was Pete Walsh, a lean man with a hawkish face, a dark outdoor tan, a stubble beard, and a headful of curly brown hair that, despite some graying, makes him look younger and more urban-hip than he is. We met at a café in the Paradise Valley, through which the Yellowstone River runs north after leaving the park. Coffee for me, tea for Nelson; Walsh wanted only a glass of water. His eyes appeared wary, reflective, a little sad, and in his frayed brown anorak he looked like a Dustin Hoffman left out in the weather, for months, to shrink and cure.

Walsh is a trapper by profession, not for amusement. "There's your average redneck, who'll just shoot a wolf 360 days a year," he said. Shoot one on sight, for the bloody hell of it, he meant—as a purblind Montana man had in fact shot wolf #10, the historic mate of that historic great-great-grandmother, soon after the male strayed beyond Yellowstone's boundary. "I've never done that, and I never

will do that." He noted that wolves, for all the angry resistance toward them, are "here to stay."

Walsh had begun trapping muskrats at age nine, near his home in central Indiana. He discovered Montana during summer visits to his grandparents but remained in the Midwest through high school, washing dishes for his tuition at a Jesuit academy. Then back to Montana for college, and he had been here ever since, except for travels to Alaska. Although he seemed almost too feral and skittish for family life, he had raised three kids, he said, mostly on his earnings from trapping. "A bobcat is a house payment." Coyotes too are lucrative. After wolf reintroduction in the park and their spillover into Montana, he began catching the occasional one in his coyote traps. They weren't yet legal game, so he released them, carefully and unharmed—eventually released 18 or 19 wolves, he reckoned. Then in 2011 they became legal for a trapper, and he had taken three so far this year. His wolf pelts, Nelson interjected, are beautifully prepared.

"I kinda love wolves," said Pete Walsh. "And I feel sorry for them. 'Cause they got put in these places where they don't belong." Of his own wolf trapping he said, and repeated, "It's a labor of love."

From the café I drove Walsh back to his truck, parked near the entrance to a ranch where he works as manager. There we sat talking further, and I heard about the Jesuit high school and the dishes. I learned that, like certain other solitary mountain folk of my acquaintance, he's a voracious reader. I heard that his daughters and son, raised on pelt sales, disapprove of his trapping, and that he approves of their right to disapprove. After kids reach the age of 11 or 12, he said, a parent figures out that they aren't marionettes.

Some people at the other pole of the spectrum, perhaps Elli Radinger and her German clients, thoughtful conservationists and animal lovers, would find Pete Walsh appalling and hypocritical. But he's a fastidious and principled man, and the paradox he embodies is not merely his own. It's built into the Yellowstone enterprise.

The Island Dilemma

Yellowstone National Park is not an island," Rick Reese wrote in that 1984 book on the Greater Yellowstone Ecosystem. He said it several times, making his case for recognizing the GYE as a larger and more integral entity than the park proper. And he was right. But it's important also to recognize that the Greater Yellowstone Ecosystem, if not the park, *is* an island in many respects—an ecological island, surrounded by a sea of human impact. It's isolated landscape. Ravens and eagles may come and go at will, but crossing from the GYE to safe habitat elsewhere is far more problematic for the likes of grizzly bears, elk, and bison. When they step off the island, they generally die.

The significance of this insularity emerged only after a revolution in ecological thinking that began quietly in the 1960s. Two scientists, Robert H. MacArthur at Princeton and Edward O. Wilson, then a young entomologist at Harvard, published a paper in 1963, then a book in 1967, the latter titled *The Theory of Island Biogeography*. The book was a drab-covered monograph that passed unnoticed by the general public but achieved vast influence in the conservation battles that followed for decades. Biogeography is the study of which creatures live where, which don't, and why. MacArthur and Wilson invoked Darwin, the Galápagos, and a long tradition since. "By studying clusters of islands," they wrote in their first paragraph, "biologists view a simpler microcosm of the seemingly infinite complexity of continental and oceanic biogeography." And insularity, they noted, is not just a characteristic of literal islands surrounded by water. "Consider, for example, the insular nature of streams, caves, gallery forest, tide pools, taiga as it breaks up in tundra, and tundra as it breaks up in taiga. The same principles apply, and will apply to an accelerating extent in the future, to formerly continuous natural habitats now being broken up by the encroachment of civilization." For instance, national parks, which they might have mentioned but didn't—not quite.

Biologists following after them defined two types of island that serve as models for the way insularity functions in the assembly and decay of ecosystems: oceanic islands and land-bridge islands. An oceanic island is one that, like the Hawaiian Islands or the Galápagos, rises up off the seafloor as molten lava and protrudes into daylight, initially sterile. Its plants, animals, fungi, and other creatures arrive by dispersal across the wide waters. A land-bridge island, in contrast, is one that starts as a teeming piece of the mainland. It once formed a knob on a peninsula attached to a continent, like Santa Catalina just off the shore of Southern California, or Tasmania off the southern coast of mainland

Australia. Land-bridge islands were connected to broader landscapes—by land bridges—until rising sea levels (at the end of the last ice age, for instance) swamped the bridges and insularized the knobs. Land-bridge islands therefore begin their insular existence with a richness of biological diversity, fully sampling the continental diversity, in contrast to the initial sterility of oceanic islands.

The central insight of MacArthur and Wilson's book was that any island, based on its sheer size (and, as later realized, its habitat diversity), could support only so much faunal diversity. An oceanic island would gain diversity, by colonization, to the point of some "equilibrium" level commensurate with its size. A land-bridge island would lose diversity, by local extinction of populations, until it declined to that "equilibrium" level. Why would it lose diversity? Because islands by definition comprise relatively small, delimited areas, which therefore support only small populations of living creatures; and small populations are more prone to being snuffed out by a variety of final causes, such as disease, hard weather, the problems of inbreeding, or bad luck. Among the famous cases of a land-bridge island losing diversity was Barro Colorado, once a hilltop in Panama, which became insularized in 1913 by the rising waters of a man-made lake built in conjunction with the Panama Canal. As a hilltop, Barro Colorado had been part of a highly diverse tropical forest, but as an island it was too small to support lasting populations of many creatures. By

The east side of the Teton Range often yields moody atmospherics whenever new weather fronts move through.

1970, it had lost the puma, the jaguar, and more than a dozen kinds of forest birds.

Jared Diamond, long before he achieved fame for his 1997 book *Guns, Germs, and Steel*, was among those ecologists who turned to the island concept for explanations and predictions of lost diversity in insularized nature reserves. Diamond's starkest and most controversial paper on this subject was "The Island Dilemma," published in 1975. It argued, among other points, that a single, large nature reserve would almost always contain more species than several small reserves of equivalent total area, simply because of island effects: Big is good, small is bad. Some other ecologists, highly cognizant that habitat diversity can be as important as sheer area, disagreed strongly. Daniel Simberloff, a former doctoral student of E. O. Wilson's and by then a professor himself, led the opposition to Diamond.

But the idea of insularity was in play. What it meant for Yellowstone, and for other national parks, was explored in the 1980s by William D. Newmark, a student in wildland management at the University of Michigan. While researching his master's thesis, Newmark discovered that the Yellowstone archives contained records of animal sightings dating back to early years in the park's history. These records told which animals had been present and, by omission, which had become absent at different points in time. Beginning work on his Ph.D., Newmark remembered those records, as well as what he had heard about MacArthur and Wilson's theory of island biogeography. Gathering similar records from other sites, he focused his dissertation on the implications of island theory for the persistence of wildlife diversity in North America's national parks. Plotting the data, he found a simple, dramatic pattern, similar to what Diamond had argued: Bigger is better.

Large national parks and park complexes had retained more kinds of mammals over time than had small ones. Most of those local extinctions resulted not from direct human persecution—as the wolves of Yellowstone had been persecuted to oblivion—but from the natural processes of extinction characteristic of islands: When habitat is constrained as a small area, animal populations remain small, and small populations tend to wink out, over time, owing to accidental factors such as disease, fire, hard weather, and bad luck. Greater Yellowstone had lost less of its mammal diversity, by such natural attrition, than had small national parks such as Zion (Utah), Bryce Canyon (Utah), and Mount Rainier (Washington State). Its size, evidently, had served it well. Newmark's findings were potent enough to merit publication by the prestigious British journal *Nature*, in a 1987 paper titled "A Land-Bridge Island Perspective on Mammalian Extinctions in Western North American National Parks."

Newmark's original work has been challenged in some particulars in the decades since, but its basic conclusion remains sound: Size does matter. The size of the Yellowstone complex helped preserve such big, fearsome, wide-ranging, combative animals as the grizzly, each individual of which demands a large territory. No other park in the lower 48 states, apart from Glacier Park along Montana's Canadian border, now supports sizable populations of the three greatest living North American carnivores—the grizzly, the wolf, the mountain lion—as well as such other predacious animals as the wolverine, the coyote, the bobcat, and the red fox. Yellowstone is the wildest park in the American heartland in part because it is our biggest, and because the Greater Yellowstone Ecosystem is much bigger still.

The other good thing about geographical bigness is that, besides giving space to large predators with broad territorial needs, it usually encompasses habitat diversity (Simberloff's concern) as well as sheer space, thereby sheltering a greater variety of creatures at all levels on the food chain, living all modes of life. That truth was reaffirmed to me by an elk hunter one December morning in Jackson Hole, Wyoming.

This hunter had killed an elk on the National Elk Refuge, which is legal by special permit (though, given the name of the place, paradoxical again) and under conditional stipulations, such as only "limited range weapons" allowed on the South Unit of the refuge. "Limited range" means muzzle-loaders or bows and arrows or other old-fashioned weapons, so as to demand more of the hunters and give an element of fair chase. When I spotted this fellow, from a nearby road, his dead elk lay on a one-wheeled game cart (provided by the refuge), and with two friends he was rolling it slowly across the bottomland toward his truck. With permission from my refuge guide, I went striding out to talk with him—committing exactly the sort of nosy intrusion that hunters with freshly killed animals seldom welcome. Once I had introduced myself and explained the basis of my nosiness, he answered my questions genially. His name was Mitch Bock, he said, "like the beer, not the composer." He lived in Fort Collins, Colorado. He had gotten his elk—a nice cow, six or eight years old, just the sort that the refuge managers hoped to see taken—with a black-powder rifle. He had killed a cow yesterday too, under another permit, and that one had required a four-hour belly-crawl through the soggy meadowland to get near her.

I thanked him for the information and was about to leave. But wait, he said, you're from *National Geographic*? Yes. And you all are doing a special project on Greater Yellowstone? Yes.

"Don't forget the little boreal toad," he said. It turned out Mitch Bock was a biologist himself, one who cared about small creatures and ecosystems, not just a sportsman who relished elk meat. The boreal toad *(Bufo boreas boreas)* is native to Yellowstone, he explained, but like so many amphibians, it isn't doing well in the modern age.

Follow the Meat

Changes in the ecosystem can be subtle but cumulative, whether caused by the detrimental factors Dave Hallac listed or by others even harder to trace. But the duties of park biologists include monitoring such changes wherever possible. That's why Kerry Gunther and Doug Smith set off on skis from a trailhead in the south-central sector of Yellowstone, on a morning in April, toward Heart Lake. There was no parking lot, no crowds, none of the famous sites or attractions nearby, just a turnout along a two-lane road amid timber; few visitors drive this road, unless they are entering or leaving the park through its south entrance. For Gunther and Smith, though, it was the takeoff point for an annual "carcass survey" to Heart Lake and its environs. Their mission was to document ungulate mortality, from both winterkill and predation, on the evidence of carcasses and other signs and to measure the use of those carcasses by grizzlies, wolves, and other carnivores. In rock-and-roll reporting the byword may be "Follow the money," but in ecology it's "Follow the meat."

Kerry Gunther has surveyed Heart Lake each year since 1992 (Smith began joining him in 1997), and other biologists did it for a decade before that, so the accumulated data represent a valuable portrait of one area through time. On this trip Gunther and Smith were slowed by two hindrances, one hinting at an ominous larger trend (climate change) and one merely a trivial inconvenience: bad snowpack and me.

The snow was crusty at first, through a thickly wooded stretch, then softer as the morning warmed and shallower as we ascended the gentle rim of the old caldera. I did my best, on creaky knees, to keep up. Near the top of Paycheck Pass, about four miles in, after crossing too many patches of gravel and mud, we took off the ski gear and paused for lunch. Then we cached skis and ski boots in some trees, hoping they wouldn't get chewed up by porcupines, and hiked onward in other boots. We left the trail and circled out toward the head of Witch Creek, a classic Yellowstone thermal area of burbling hot spots and warm springs, which stays snow-free and green during even a harsh winter. There we found thickets of dead bracken fern, wide-leaf grasses, mosses, liverworts, and crusty soil showing signs of sulfurous venting. Gunther spotted a grizzly print in dried clay, measured it, took GPS coordinates, and recorded the data on his clipboard. Then we went on, circling beneath the snowy glower of Mount Sheridan (named for General Phil) toward the lake.

Before wolf reintroduction, this clement drainage supported 40 or 50 elk through each winter. A few would die of starvation if not cold, and the local grizzly bears, emerging in spring from their dens, would feast on those carcasses. That much had been documented by earlier carcass surveys before the Gunther-Smith era. But the dynamics changed in the late 1990s, when the wolf reentered the picture. Dispersing down from the Lamar release points within just a couple years, venturesome wolves discovered easy pickings at Witch Creek. They killed some of the elk, probably by driving them out of the warm zone into deep snow, as wolves do, and scared the others away. Nowadays no elk winter in the little basin, and therefore no elk carcasses lie available when the grizzlies awake, groggy and famished from their long hibernation. But the odd thing, Gunther told me, is that Witch Creek's grizzlies have remained. The elk are gone; the wolves are gone. But the big bears seem to like the area for its remoteness and security.

What do they eat in spring?

"Earthworms," he said. Earthworms plus a few other items—including pocket gophers and spring beauty wildflowers and ants. That was the simple version of a complicated story to which he would return later in our trip.

We made camp in the Heart Lake Ranger Cabin, which in past years would have been bunkered with four feet of snow into early May but this year was clear. Instead of shoveling it out, we basked on the front porch, drinking beer (from a cache in the root cellar) and admiring the view. Next morning we hiked the soggy west side of the lake, Gunther scouting for grizzly tracks and Smith, whose portfolio includes birds as well as wolves, taking note of each flicker, junco, and sapsucker. On the return, we traversed higher upslope to Rustic Geyser, an oval pool 30 feet wide and filled with steaming turquoise water. In the deep end lay the skull and leg bones of an elk that had blundered to its death. Gunther recorded some bear scat nearby and then drew my attention to large clumps of sod that had been ripped out of a wet, grassy bank. "This was an earthworm dig," he said. "It's kind of amazing, something as big as a grizzly bear, making a living from earthworms." The bears rely on this resource only briefly, he explained, while spring snowmelt saturating the soil drives the worms up near the surface. They might also excavate gophers, eat the sugary corms of spring beauties, and consume quantities of clover and grass. A month later, when elk return here for calving, the grizzlies would prey on their calves.

Another irony, he added, is that these earthworms are exotics, not members of native species. (My later research confirmed it: After the last ice age, native earthworms were absent from northern North America, including the Yellowstone area, and into that absence came European earthworms, transported inadvertently by people bringing over their preferred old-world plants in soil. The European worms thrived and became dominant, as European humans did, in the new world.) "Not all exotics are necessarily bad," Gunther added. Soil

Bocats usually prowl furtively, though in recent years a family has become visible along the Madison River in Yellowstone.

scientists could disagree—the Euro worms have altered nutrient cycles—but he was speaking as a bear biologist.

"In this case, we've got exotic clover, we've got exotic earthworms, and the bears benefit from it." He grinned wincingly, mindful of all the continual fixing and jiggering demanded of Yellowstone managers. "And we have no plans to try to eradicate earthworms."

I suppose managing nature is an imprecise science, I said. You've got to deal with the facts on the ground. And history.

"And history," Gunther agreed.

Late that afternoon, after we had rinsed off our sweat in the ice-free end of Heart Lake, Doug Smith described another cascade of consequences involving invasive species in Yellowstone. The lake trout in Yellowstone Lake have caused trouble for trumpeter swans, loons, and osprey, as well as for grizzlies. Osprey are specialist raptors that eat only fish, and on Yellowstone Lake that has meant cutthroat trout, which feed on insects near the surface and make themselves vulnerable to the birds, swooping low with ready talons. Now the cutthroat are nearly gone and so are most of the osprey; from about 40 breeding pairs, the lake population has fallen to just 4 osprey pairs.

Bald eagles eat fish too, but bald eagles are generalists, with two other categories of customary prey: carrion and waterfowl. Deprived of cutthroat, the bald eagles from Yellowstone Lake have begun ranging more widely, Smith explained, and coming down hard on some of our most valued waterbirds. Loons, for instance. The common loon is abundant in Canada and Alaska, but our Yellowstone population represents a Pleistocene relict, meaning that it's isolated at the southern extreme

of the range, left behind atop this plateau after the ice receded, and therefore of special interest. Right here at Heart Lake, Smith said, we formerly had two nesting pairs of loons. "There are none now. Zero."

As for trumpeter swans, they had their own troubles, he explained, even before bald eagles started killing them. The trumpeter is the heftiest North American bird, a majestic thing with a 10-foot wingspan, but sensitive to disruption, especially as cygnets (chicks), which sometimes face high mortality from predation, nest flooding at the egg stage, and starvation. A century ago, the adults suffered heavy hunting for their luscious skins and feathers. By the 1930s, trumpeters had been nearly extirpated from the lower 48 states, with the Yellowstone ecosystem providing their only stronghold south of Canada. Then they were reintroduced elsewhere and made a modest comeback, but the population in Yellowstone Park has declined since the 1960s. "We went from about fifteen swan territories down to two," Smith said. One of those two breeding pairs is at Riddle Lake, a small body of water not far from Yellowstone Lake and surrounded by good habitat for grizzly bears in early summer. Part of Riddle's shoreline is thick with vegetation, which provides cover for nesting swans and their newly hatched cygnets.

Beginning in 2010, the Yellowstone swans failed to fledge any new cygnets for several years, and Smith wondered why. He passed over Riddle Lake on a survey flight (with Roger Stradley at the stick) and saw two adult swans in the middle. "An odd place to be, because it's open water and there's no cover." Then he spotted a dead cygnet floating upside down, its feathers strewn all around. Perched in a tree nearby, drying its wings, was a bald eagle. "Oh my god, this is a smoking gun," Smith thought. The eagle had just killed the cygnet, evidently, and was waiting for the adults to move before claiming its prey. But why had the swans all exposed themselves in mid-lake?

Smith and other researchers pieced the story together. Because of its heavy use by bears, Riddle Lake was closed to visitors until July 15 each year. That gave privacy also to the trumpeter swans, at least during the early weeks of their nesting and rearing cycle. But when the trail opened in mid-July and hikers arrived, human ruckus drove the swans from their safe zone along the shore. That helped the eagles, Smith said, and "they were killing these swan cygnets every year."

So the administrative closure of Riddle Lake was extended to September 1, prompting angry calls from thwarted hikers who complained that park management was too restrictive, infringing on their use of Yellowstone as a "pleasuring-ground" for the "enjoyment of the people," as stipulated in the act of establishment. Smith and colleagues had meanwhile begun supplementing the swan population with trumpeters hatched and reared in captivity, "grafting" them back into the wild at one day old. "This is radical for the park," he admitted, just as culling of bison is, at least since the Leopold Report and the espousal of "natural regulation." But even the longer closure of Riddle Lake hasn't solved the eagle problem, Smith told me. When the lake trail came open on September 1, and hikers arrived, "*Bam*, eagle flies in and nails both cygnets." The lake is now closed until September 15.

"We're fighting against this kind of public tide of 'What about us?'" he said, and further paraphrased the voices of complaint: "*It's our park, we're taxpayers, we want to enjoy it.*" That tide has ebbed and flowed

A beaver dives through a side channel of the Snake River. Trapped out by fur traders in the 19th century, beavers are rebounding in number across Greater Yellowstone.

CHARLIE HAMILTON JAMES

throughout the history of the park, but its amplitude has increased with the rise of annual visitor numbers into the multiple millions.

Listening to these two men discuss such concerns, in the tranquil setting of Heart Lake, was like a seminar in the difficulties of park management. Smith with his trumpeter swans, his marauding eagles, his loons, his rankled hikers, and Gunther with his grizzly bears eating European earthworms, reminded me of what makes their jobs so tricky, their work so worthy, and their chances of final success, to the full satisfaction of all parties, so remote: It's the paradox of the cultivated wild. ■

*To pinpoint the cor-
ridors that migratory
elk use through the
mountains near the
southeast corner of
Yellowstone, a hardy
researcher shadows
a herd that rumbled
through a few hours
before.*

JOE RIIS

Hurling herself up and over, a female coyote long-jumps toward a snow tunnel where a vole is trying to escape. This hunting expedition in Yellowstone's Hayden Valley near Mud Volcano produced 15 voles for the coyote in less than 10 minutes.

RONAN DONOVAN

It can be a brutal, hard-knocks life, even for a wolf. Here, in Pelican Valley, a member of the Mollie wolf pack inspects the remains of an elk bull that a few days earlier had killed the pack's alpha male as the wolves tried to bring down the wapiti.

RONAN DONOVAN

(opposite)
Yellowstone's senior wolf biologist Doug Smith targets a member of the Mollie pack with a tranquilizer dart — part of a research effort to learn more about wolf behavior.

DAVID GUTTENFELDER

Meanwhile, at another location, a large male member of the Eight Mile pack has been temporarily tranquilized to collar for monitoring.

RONAN DONOVAN

(following pages)
An ice-sheathed waterfall roars into the chasm known as the Grand Canyon of the Yellowstone. The Yellowstone River, which flows through it, is the longest undammed wild river in the lower 48.

MICHAEL NICHOLS

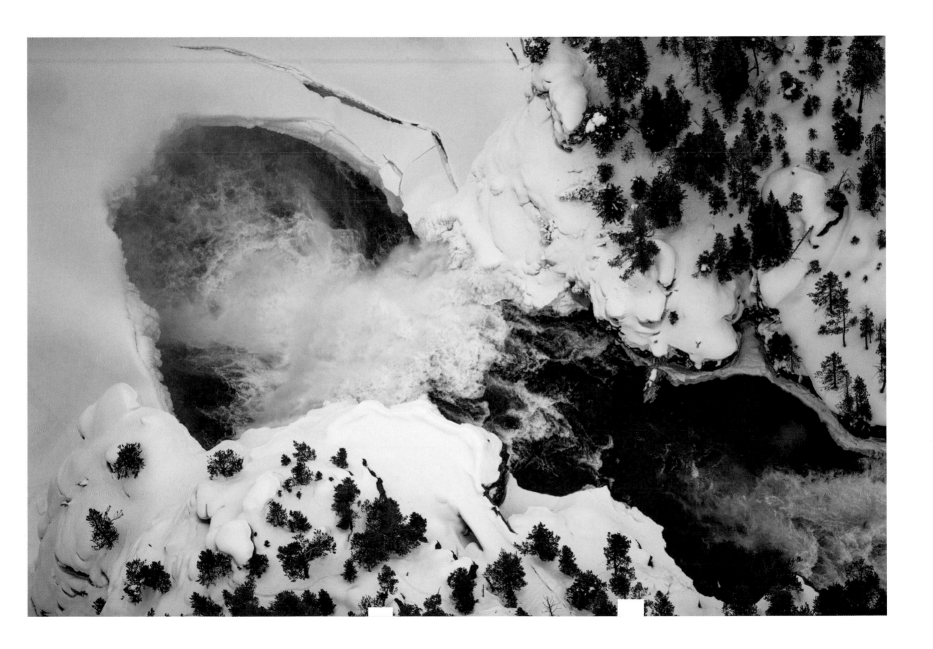

Detail

Migrations

Seasonal migrations follow paths carved into the landscape long before humans drew in boundaries of ownership or protection.

In spring hoofbeats return to the mountains of the Greater Yellowstone Ecosystem. Pronghorn, North America's land-speed champions, migrate more than a hundred miles from the Green River Valley into Grand Teton National Park. Mule deer mosey 150 miles north from the Red Desert Basin to the mountains around Hoback Basin, just south of the park. Herds of elk stream into Yellowstone Park along a web of migration routes; in October they fan back out again. Tanzania's Serengeti Plain, with its thunderous migrations, comes to mind—except these are happening in the western United States, in an expanding modern economy. And that's the challenge.

The circulation of the ungulates is like a heartbeat that pumps life into the ecosystem; it "gives animation to the spirit of this amazing place," says Arthur Middleton, an ecologist at Yale. "But there is a risk of cardiac arrest." If the migration corridors are the arteries of the ecosystem, he adds, in some cases they're being constricted and blocked by myriad human developments: oil fields, subdivisions, highways, fences. An initiative to preserve the migratory path of the pronghorn offers hope by uniting government land managers, landowners, conservationists, and hunters. Middleton and his colleagues are compiling an atlas of migrations to guide policymakers. All share the aim of allowing pronghorn to follow the paths their ancestors have followed for millennia—and even cut loose at 55 miles an hour now and then.

New research on elk migration has revealed the truly amazing and breathtaking lengths that wapiti go to to pass between winter and summer range—journeying literally up and over mountains.

JOE RIIS

National Geographic photographer Joe Riis, working with scientist Arthur Middleton and a team of colleagues, has chronicled the dramatic migrations of elk. They ford raging rivers (opposite), such as the South Fork of the Shoshone River, at 6,000 feet in elevation, and return from summer grasses to pass along the edge of steep mountain cliffs (right), near the Thorofare Plateau.

A bald eagle helps itself to the carcass of a pronghorn (antelope) that died while trying to swim the Upper Green River.

JOE RIIS

South of Yellowstone, pronghorn migrate seasonally, spending summers in Grand Teton National Park and winters in the Upper Green River Basin, more than 100 miles away. The corridor is called Path of the Pronghorn.

JOE RIIS

(opposite)
The migratory paths are part of ancient behavior passed along through generations—and information about these migrations is essential to discussions on how to protect this aspect of wildness in Greater Yellowstone.

JOE RIIS

If a migratory corridor becomes disrupted, pronghorn trying to leave Grand Teton Park during the winter could get stranded and die. Seemingly simple things, such as restringing fences with wildlife-friendly line, can have huge positive consequences.

JOE RIIS

Where abundant prey resides, so too do thriving predator populations. Predators such as wolves cull wild herds and thus help slow the progress of wildlife diseases.

JOE RIIS

(following pages, left)
With snow-dappled antlers, a bull elk crosses a high divide near the Thorofare Plateau at night, notably near the same spot where a camera trap documented a nomadic grizzly bear.

JOE RIIS

(following pages, right)
Notoriously cryptic, cougars vary their range in response to prey. In winter, the tawny-colored cats favor the shallow snow in the northern reaches of Yellowstone.

DREW RUSH WITH THE
NATIONAL PARK SERVICE

Bill Hoppe

RANCHER, WOLF-RELEASE CRITIC

"My great-grandfather was the first white man born in the Montana Territory. We have always lived in this area, even before there was a national park. We love this area, and we love the park. It is really when they started calling it an ecosystem that all the problems began. People from all over the world having an opinion on how this area should be managed and how we should be ranching, hunting, and living our lives. People that have never once had a grizzly at their front door when their wife walks out to go to work in the morning."

Bill Hoppe has deep connections to his homeland on the edge of Yellowstone, and he's not happy about outsiders preaching how it ought to be managed.

ERIKA LARSEN

PART FOUR

The Future of the Wild

There is one spot left, a single rock about which this tide will break, and past which it will sweep, leaving it undefiled by the unsightly traces of civilization. Here in this Yellowstone Park the large game of the West may be preserved from extermination.

—GEORGE BIRD GRINNELL, 1882

Hillary Anderson cuts a mythical cowgirl figure for the 21st century as she rides the range above her father-in-law's ranch in the Tom Miner Basin, north of Yellowstone.

CORY RICHARDS

Dem Bones

Everything is connected. That's the first lesson not just of ecology but also of resource politics.

The wolf is connected to the grizzly bear by way of their competition for ungulate prey, especially elk calves and adult elk that have been weakened by winter or the rigors of the autumn rut. Whitebark pines are connected to mountain pine beetles, whose population outbreaks are connected to climate change. Bison are connected to Montana livestock policy by way of a disease, brucellosis, brought to America by cattle. Elk are connected to the boreal toad by way of Mitch Bock.

Elk are also connected to cutthroat trout. In this case it's by way of grizzly predation, taking a bigger toll on elk—some evidence suggests—since the crash of the Yellowstone Lake cutthroat. Aspen trees and willows are connected to both wolves and grizzlies by way of the heavier elk predation, which does seem to have helped aspen and willow stands recover (notwithstanding Arthur Middleton's scorn for the simplistic version of this story) from decades of heavy browsing. Then again, decades of drought and the absence of beaver, whose "habitat engineering" raises the water table, may also have contributed to the historic aspen and willow suppression. Trumpeter swans are clearly connected to cutthroat trout by way of bald eagles. Moose are connected to beaver because moose eat willows, but they're connected also to mule deer by way of a parasitic nematode worm (*Elaeophora schneideri*), for which mule deer are the usual host. Horseflies carry the worm from deer to deer—or from deer to moose, brokering that connection. The worm is innocuous in mule deer, but in moose it restricts blood flow to the head, sometimes causing brain damage, blindness, and death. Grizzly bears

---

Most bison alive today in Greater Yellowstone can trace their origins to a couple of dozen individuals that found refuge, like this one still does, in the center of Yellowstone.

are connected to corn farmers in Kansas, and their decisions on pesticide use, by way of army cutworm moths.

The changes that ricochet through these networks of connection, from animal to plant, predator to prey, one level of the food web to another, are known to ecologists as "trophic cascades." They are a focus of interest, and disagreement, among scientists who study the wildlife and vegetation of Yellowstone, including Doug Smith and Arthur Middleton. The details of those disagreements become almost Talmudic in complexity, but what's important to keep in mind is that disturbances have secondary effects, usually unforeseen, and that sometimes those effects are irreversible. Restoring wolves to Yellowstone, for instance, does not necessarily fix all the problems that removing wolves from Yellowstone caused.

Such interconnections underscore the truth of a truism: that the Greater Yellowstone Ecosystem is an intricate, interactive compoundment of living creatures, relationships, physical factors, geological circumstances, historical accidents, and biological processes. As an ecosystem, in all its glories and its troubles, its fractured relationships and the consequences of those fractures, it can teach us a lot about how nature works. Its greatest value, its fullest purpose, is not simply to freeze a picturesque place "in the condition that prevailed when the area was first visited by the white man," as the Leopold Report proposed. Why measure time from the arrival of John Colter, Jim Bridger, and Nathaniel Langford? Ecosystems are continually subject to change, both from human-delivered disturbances (including those caused by indigenous peoples before "the white man" arrived) and from natural ones. The long outline of the Yellowstone caldera, nicely visible from the top of Mount Washburn, should be enough to remind us of that. In its aboriginal condition as of 640,000 years ago, the whole place was just a vast smoking hole.

Instead of a smoking hole, we wanted a landscape full of living creatures as well as geysers and canyons, and by a long series of visionary acts, heroic efforts, mistakes, corrections, happy accidents, and good decisions, we have it in 2016. Superintendent Dan Wenk told me that he thinks Yellowstone National Park, for all its problems, might be in better overall shape now than at any time since 1975 (the year the grizzly bear's decline was recognized with federal protection). Dave Hallac, notwithstanding his concern with the creeping crisis, agreed.

Whether the same can be said of the Greater Yellowstone Ecosystem—that it's in better shape now than for many decades past—is more questionable. Have we vastly improved this great area from the bad old days of commercial poaching and vandalism, governmental neglect, Wild West brigandage and railroad dreams, to the present good moment—or have we already gone a long way toward making it a big, boring suburb with antler-motif doorknobs?

The GYE is a focus of many angers and worries, in part because at its core lie two precious national parks, surrounded by other lands, public and private, which are subject to different expectations and governed by different interests. Some hunters are angry that there aren't enough elk. Some ranchers are angry that there are too many. Some wolf lovers are angry that wolves, including those that spend much of their year within Yellowstone Park, may now be hunted or trapped when they roam beyond the park boundaries. Some land-owners in Gardiner, Montana, are angry that bison migrate out of the park, in winter, and into their yards. Some stockmen are angry that migrating bison carry brucellosis, which might be passed to their cows, although not a single case of bison-to-cow transmission has ever been documented. Some wildlife activists, including those of the Buffalo Field Campaign, are angry that bison from the park, as they migrate out, may be corralled and shipped to slaughter. Some range scientists are angry about overgrazed grasslands in the two parks, resulting from too many bison and elk. Some fishermen are angry about the slaughter of lake trout. Somebody somewhere is probably angry about coyotes. Scarcely a season passes, in the gateway towns of Cody and Jackson and Bozeman, without several public meetings, called by the various agencies, at which people express these angers.

One such meeting occurred last December in a hotel conference room at Jackson's ski resort. Roughly a hundred people crowded in, interested citizens filling rows of chairs, some standing at the back, to hear scientists and managers deliver updates to an interagency committee charged with oversight of the Yellowstone grizzly bear. The atmosphere was tense and adversarial. Many people in the room had fought one another over this issue for decades. The crowd heard Chris Servheen, coordinator of grizzly recovery for the U.S. Fish and Wildlife Service (a position he has held for 35 years), explain that the Yellowstone population has now reached its benchmarks—"We consider the bear recovered"—and that the FWS would soon propose removing it from listing as threatened under the Endangered Species Act.

How soon? Very, but indefinite. (In fact the proposal came through several months later, in early spring of 2016, while this book was in final preparation, and it still remains to be reviewed by the public and acted upon during the forthcoming year.) Such delisting is contentious because, once it happens, the states of Wyoming, Montana, and Idaho will be free to issue licenses to hunt grizzlies, for the first time since 1975.

Frank van Manen, leader of the Interagency Grizzly Bear Study Team, made a dispassionate presentation, which included encouraging data on current grizzly numbers and their dispersal throughout the ecosystem, and Servheen discussed those data, all to a largely skeptical crowd. Other scientists and managers then spoke, answered polite questions from the committee, edgier ones from the audience. Toward the end of the afternoon, the floor was opened for public comment from any people who had entered their names on a list.

A man named Reuben Fast Horse, representing the Oglala Sioux Tribe, walked to the podium carrying a small drum. His longish dark hair was pushed back behind his ears. He adjusted the microphone, took off his glasses, and began speaking in Lakota, one of the Sioux languages. All the white people listened raptly. Occasionally we caught a word—"Europeans." Fast Horse spoke for three minutes, his speech musical but utterly incomprehensible to most of us, then came to an end point, and said: "Don't worry, the rest will be in English."

He read a statement from the president of the tribe expressing strong opposition by the Oglala Sioux Tribal Council to delisting the grizzly. This statement noted a relationship with the bear "that has existed from time immemorial" and stressed that the tribe "recognizes the grizzly as a relative, a healer and a teacher of our people, as exemplified in narratives related to our ancestors." Finished reading, Fast Horse added some words of his own, explaining that his Lakota people "never ate or hunted bear" anymore because, when they originally had, "the skeletal remains looked too human, too close to ourselves." Then he lifted his drum and his stick, played a strong cadence, and sang a bear song in Lakota.

Imagine you're a federal manager, like Dan Wenk, seated at the committee table. How do you reconcile *that* with the metrics of population biology?

Wenk himself has concerns about grizzly delisting—not that the bear population hasn't strongly recovered, but that the negotiating process doesn't directly include the National Park Service. That process is "owned by the Fish and Wildlife Service," he told me later, with park voices left "peripheral." Will the new management regime weigh the interests of Yellowstone and Grand Teton visitors, who want to see bears, equally with the interests of hunters? Will the economics of bear-watching be duly considered? What happens if a grizzly is wounded in Montana, then comes roaring back across the boundary to suffer its agonies in Yellowstone? What about visitor safety? What about perceptions? Who puts the animal out of its misery? Who takes the heat? Does that bear count in the hunting limits?

I asked Wenk whether he had gotten satisfactory answers to these questions. He's a patient man, and a professional. He said: "Not yet."

Ownership

Wenk's concern reflects an important truth: that the people who live and work and hunt and fish and hike within the Greater Yellowstone Ecosystem— even people such as Reuben Fast Horse, whose local ancestry goes back millennia—are not the sole possessors of legitimate interest. This is America's place, and the world's. Animal lovers at the far reaches of Twitter, people who have never given 10 minutes' thought to grizzly bear conservation, who could never tell you the Lakota word for grizzly or the bear's four major foods in Yellowstone, are angry at Superintendent Wenk for ordering the death of the sow that killed Lance Crosby in 2015. Yellowstone National Park received four million visitors that year; Grand Teton National Park (GTNP) received more than three million—and visitors feel invested, which is good. As the superintendent of Grand Teton, David Vela, said last July to a group of Latino schoolkids from Jackson who were spending a week in GTNP as part of an outreach program: "You own this national park. This is part of your heritage as Americans."

Wes Livingston, the brusque and candid mountain guide, the inveterate hunter and antler collector, the detester of government regulation, the despiser of fatuous liberals and whiney cattlemen, captured much

With livestock grazing the slopes and a country road running through, the foothills of the Absaroka Range north of Yellowstone represent a dividing line between civilization and wildness.

TOM MURPHY

the same spirit one evening in camp in the Thorofare. Livingston views Yellowstone Park as a single big ranch, managed for desirable animals. In other words: the cultivated wild. Who owns this ranch? I asked.

"We do," he said.

"Who is 'we'?"

"The United States of America. The citizens of America, the taxpayers."

"Does an outfitter in Cody own it more than a wolf hugger in New Jersey?"

"Absolutely not," Wes said.

Other issues stir quiet worries. Agency biologists such as Mark Bruscino worry that the loss of big private ranches to subdivision will destroy migration routes and winter range of public wildlife. Grizzly bear advocates such as Doug Peacock (author of *Grizzly Years* and a legendary figure in the West) and Dave Mattson (longtime bear scientist and former member of the Interagency Grizzly Bear Study Team) worry that, with some of the major foods lost, delisting and the hunting to follow will doom the Yellowstone grizzly. Others worry that *failure* to delist the bear, given its robust population recovery, will only further inflame resentment against the grizzly among people with whom it shares habitat and will undermine the Endangered Species Act itself. Bird lovers worry that the trumpeter swan may be eradicated from Yellowstone. Wildlife veterinarians worry about the approach of chronic wasting disease, a bizarre affliction similar to mad cow disease, spreading northward toward the Greater Yellowstone Ecosystem among mule deer. Dave Hallac worries that climate change, population growth, private lands development, lack of coordinated management, and other factors will creep across Yellowstone and slowly ruin it. Rangers worry that still another clueless tourist will be gored while taking a selfie in front of a bison.

As if that weren't enough, some people worry (despite reassurances from experts such as Robert B. Smith) that the Yellowstone supervolcano will explode again soon, incinerating everything and everyone within hundreds of miles.

Bob Smith and his colleagues, by the way, made another spooky discovery within the past decade. They found that the Yellowstone caldera went into "a rapid episode of ground uplift" from 2004 to 2010, as revealed by their fancy measuring devices. It has been moving like the diaphragm of some gigantic creature taking big breaths. The likely explanation for these uplifts is recharging of the upper magma chamber, as new molten material ascends from below. Does it mean that the whole thing is ready to blow? Smith says no.

And me? My own worries focus on the grizzly bear, because I consider this the highest and best purpose of Yellowstone National Park and the Greater Yellowstone Ecosystem: to preserve a viable population of that great, terrible animal at the center of the American West into the indefinite future. I don't share the extreme pessimism of my friends Peacock and Mattson, nor their distrust of the conscientious agency biologists, such as Kerry Gunther and Mark Bruscino and many others I've met, who believe that the bear's intelligence and flexibility of behavior will keep it thriving and numerous despite changes in the landscape that require greater reliance on some different foods. After all, these biologists say, the grizzly is an omnivore, and three out of the four major dietary items—those three being the trout, the pine nuts, the moths—have never been available to all Yellowstone's bears every year or even to many bears during some years. Whitebark pine nuts come and go in cycles. Spawning cutthroats offered nothing to bears on the west side of Yellowstone, distant from the lake. Likewise with army cutworm moths: Many grizzlies exploit them, but not all grizzlies. Change may come, they say, but the bear will adapt to the challenge.

Still, Dan Wenk is right that the devil could be in the details, that the opaque process of delisting and the return of responsibility to the states contain possibilities of harm for the grizzlies, and for the public's interest in them, against which there must be guarantees. I share the concern of Wenk and others that the process of delisting and the management regimes under which it is implemented in the three states, when that time comes, should be fully sensitive to the interests of park visitors and other citizens as well as hunters and ranchers. It's also crucial, I think, that stipulations exist for *re*listing the grizzly immediately in the event that failing food sources, invasive species, overhunting, or any other factor tips the population back toward insecure levels.

Meanwhile, we the owners of Yellowstone National Park, Grand Teton, and the national forests and other federal lands of the ecosystem face some new challenges of our own. The parks need better funding for the impossible work they do; only a fraction of their operating and improvement funds comes from Congress, whereas crucial initiatives such as the Yellowstone Wolf Project are supported by private money, through "friends" organizations such as the Yellowstone Park Foundation. The parks also need political support for hard decisions, such as the one that may come when, because of overcrowding, private automobiles are no longer allowed to enter. Sorry: Get on the shuttle. The most heated wildlife issues, notably grizzly and bison and wolf, need collaborative solutions, not continuing warfare. Passionately dedicated people need to recognize that righteous intransigence is not a strategy; it's just a satisfying attitude. The various agency members of the Greater Yellowstone Coordinating Committee need to add private groups as partners in a more efficacious coordinating body and to make bold decisions that transcend turf politics. Climate change is hurting Yellowstone—by way of temperature ranges, insect cycles, drought, who knows what else—and we all need to do better on fixing that.

Hah, easier said than done. These are torturous challenges requiring uncomfortable, scarcely imaginable solutions. But if the Yellowstone grizzly bear is expected to adapt, to modify its behavior, and to cope with new realities, shouldn't we be expected to do that too?

The Beginning

With his elk calves counted, Arthur Middleton and I rode still higher on the Thorofare Plateau, above our camp, above the summering elk, until the rising swell of land crested as a narrow ridge at about 10,300 feet. I was amazed that backcountry horses could climb so high and keep their balance on such footing; it seemed extraordinary even for crazyass, obdurate mountain steeds trained by the Livingston

brothers; but I tried not to think about that. Gnarled whitebark pine, much of it dead but some alive, formed a thin windbreak along the ridge. We crossed rocky ground, and meadow enlivened by Indian paintbrush, and then patches of mud just melted out from beneath last winter's snow, where in a week or two fresh alpine grasses might grow. The vista around us, through 360 degrees, was epic.

I turned in my saddle, trying to see it all: the Absaroka Range to the northeast, the Trident Plateau just across Thorofare Creek, Yellowstone Lake in the distance beyond that, and even farther, the Gallatin Range. I cranked around. Westward was Two Ocean Plateau, then the Tetons to our southwest, with the Grand itself rising, unmistakable in height and profile, to 13,770 feet above the Snake River and Grand Teton National Park. Coming toward us from the south: the highest reaches of the upper Yellowstone River. Arthur pointed across the Castle Creek cirque to where, along a north-facing rim, sizable cornices and couloirs of snow still lingered, shedding their melt to the streams below. This was late July. Look at all that water, he said. It drains to the big river. It grows the grass to feed the elk.

To feed the whole thing, I thought: all the players, all the processes—the elk and the grizzly and the cutthroat trout, the photosynthesis and the herbivory and the predation, the competition and the migration, the parasitism and the decomposition, everything

A newborn elk calf, vulnerable and delicate, rests in a safe spot near Yellowstone's administrative headquarters in Mammoth Hot Springs.

MICHAEL NICHOLS

downstream, everything that moves into Yellowstone and across it and back out. Seems almost like this is where the ecosystem begins, I said.

"If an ecosystem 'begins' anywhere," he agreed, "this would be it."

For every beginning in the natural world, there is an ending, and then a beginning again. Does that robust cyclicity hold true even when nature's wildness is cultivated by people? Maybe. Maybe it does, if the cultivation is judicious, humble, and wise. We're up so high here, atop the Thorofare Plateau, you and I and Arthur Middleton, that a person can almost see the future. ∎

Hugging Old Faithful, a national natural treasure, stands a national architectural treasure: the 1920s-era complex of buildings including the Old Faithful Inn and Old Faithful Snow Lodge.

MICHAEL NICHOLS

***Local cowboys guide
visiting dudes on a
horseback ride into
Yellowstone's Lamar
Valley. It's an activity
that still endures in a
landscape more
"Wild West" in some
ways than at any
point in 140 years.***

ERIKA LARSEN

***Tumbling 150 feet
in cascading ribbons,
Dunanda Falls on
Boundary Creek
meets the twilight
with a rare "moon-
bow," an evening
equivalent of a
rainbow, formed
at its feet.***

MICHAEL NICHOLS

A glimpse of the New West: In the Gravelly Mountains of Montana, John Helle and his two sons stay with their 1,400 sheep (opposite) on public lands through the summer grazing season. A Peruvian shepherd and two Akbash guard dogs including this one (right) help tend the massive flock. Constant vigilance and bear spray replace bullets as a way of managing bears and wolves.

Employing the traditional method of hauling a harvested elk home for butchering, a hunter pulls his quarry out of the backcountry just beyond the northern boundary of Yellowstone. Elk hunting is a savored activity for Montanans and a business contributing millions of dollars to the regional economy.

DAVID GUTTENFELDER

Leo Teton

MEMBER OF THE SHOSHONE-BANNOCK TRIBES
Fort Hall Reservation, Idaho

"What does the buffalo mean to me? In the 1800s the buffalo was almost extinct due to the killing by the white men. But today we still use the buffalo in all our ceremonies. As a sweat lodge owner, I use the buffalo skull in my lodge. I place the buffalo skull in front of my doorway on a dirt mound. This altar represents strength and good, long, healthy life to all those who enter … We pray for the buffalo to have a good travel to the other side, the animal spirit world."

At Fort Hall, Idaho, outside a sweat lodge where Shoshone-Bannock tribal members gather for ritual purification ceremonies, Leo Teton stands next to a pole ornamented with bison skulls.

ERIKA LARSEN

To enter Hayden Valley in summer when bison cows and calves spread far and wide across the rolling hills, and to hear bulls snorting in the rut, kicking up dust and wallowing, is like glimpsing into the American prairie centuries ago.

MICHAEL NICHOLS

(opposite)
Bison specialist Todd Traucht inspects a Yellowstone animal before shipping it to the Fort Peck Indian Reservation in eastern Montana, where the tribe is enthusiastically building its own herd.

DAVID GUTTENFELDER

At the Fort Belknap Indian Reservation in north-central Montana, Yellowstone bison are held in a facility until they, too, can pass quarantine, proving they don't carry disease.

DAVID GUTTENFELDER

(preceding pages)
Younger generations of ranchers are learning to employ new nonlethal techniques for coexisting with grizzlies and wolves, including fladry lines strung with colorful fabric (left) as a visual deterrent to wolves. For westerners living near Yellowstone, so much is changing. Here, four-year-old Elle Anderson (right) chases a ball—and, with any luck, a meaningful future on her parents' ranch in Montana.

LOUISE JOHNS

The Thorofare region, one of Greater Yellowstone's healthy highways for elk, cradles wildlife populations unbothered by the trappings of modernity and becomes a touchstone for thinking about the meaning of progress.

JOE RIIS

The Park as Art

Thomas Moran's grand paintings of Yellowstone scenery changed the public impression of this landscape forever.

Before Thomas Moran arrived, Yellowstone in the popular imagination was a harsh, wild place pocked with hellish geysers. After the painter's work was finished, Yellowstone was a national park marketed as a wonderland.

In 1871 Moran, a young English artist, joined photographer William Henry Jackson and took part in the first U.S. government survey of the region. For two weeks Moran filled a sketchbook with the region's most stunning sights. The survey results, Jackson's photos, and Moran's watercolors—the first color representations of the area—were presented to Congress that fall as part of a proposal to create a national park. The proposal succeeded, and in March 1872 lawmakers officially established the world's first national park.

Moran's practice was to create dozens of watercolor studies and then re-create them as large oil paintings. After the 1871 expedition he did just that: By April 1872 he had transformed some of his sketches into a 7-by-12-foot painting (opposite, upper left). The gold-spattered valley and billowing Lower Falls of "The Grand Canyon of the Yellowstone" captivated the public. "It is too grand and wonderful for words," declared the Ladies' Repository that August, "and none can ever judge of its wonders from any engraving or photograph in mere black and white." The painting now hangs in the Smithsonian American Art Museum, along with several other landscapes by Moran.

He continued to travel the world and the American West, painting landscapes of Mexico, Venice, Long Island, the Grand Canyon, and many other locations. Yet his reputation was so intertwined with Yellowstone that he took to signing his paintings "TYM," for Thomas "Yellowstone" Moran.

In one of the earliest photos taken by William Henry Jackson during the 1871 Hayden Expedition, when the groundwork was laid for Yellowstone becoming a national park, master painter Thomas Moran stands on a travertine terrace at Mammoth Hot Springs.

W. H. JACKSON

The Grand Canyon of the Yellowstone, 1871

Tower Creek, 1871

Tower Falls & Sulphur Mountain, 1875

The Tetons, 1879

Still evolving, drip by drip: Palette Spring, growing on the terraces at Mammoth Hot Springs, is also perhaps a metaphor for Yellowstone as a big, bold idea that continues to take shape. Only by protecting the physical essence of Yellowstone can society also behold its intangible qualities as a public space that celebrates nature and is owned by all citizens.

TOM MURPHY

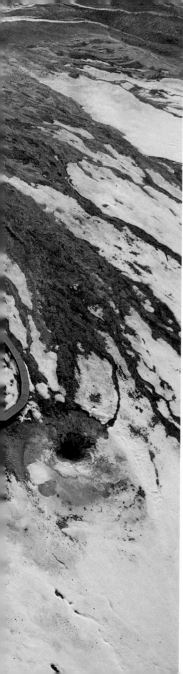

*Every year NASA
spacecraft send back
images of other plan-
ets devoid of life,
while right here, in
America's backyard,
there's an exotic land-
scape, teeming with
spectacular life forms
large and small, that
exceeds anything seen
on Mars.*

MICHAEL NICHOLS

In that perfect balanced moment between sunset and moonrise, wolves howl their haunting calls of the wild on a bluff in Greater Yellowstone.

JOE RIIS

What is wild? What is natural? What are the things in nature that possess value for society? A riding horse in Montana munches its way through a pasture abundant in both native and interloping wildflower species.

ERIKA LARSEN

The power of Yellowstone's beauty today is as enduring as when humans first spied it. Here, a large backwater pool along Alum Creek in the Hayden Valley perfectly reflects the sky and the Absaroka Range towering above.

Views of the Grand Canyon of the Yellowstone, like this one from Artist Point, never fail to elicit awe. Somehow we know that this is what "breathtaking" is supposed to be.

MICHAEL NICHOLS

Becky Weed

RANCHER, BELGRADE, MONTANA

"We can pick our poison. Castle-building landowners who are busily resurrecting a feudal society while chopping up habitat. Ranchers and politicians who are too quick to put wildlife in the crosshairs as a scapegoat for deeper ills in our agricultural economic system. Energy companies' boom-and-bust frenzies. The press of all the rest of humanity. It's hard to condemn any one sector without acknowledging the warts and complicity of any other, but collectively we're degrading the magic that makes this region unique. Can we slow down, scale back, and proceed with less of an air of entitlement?"

Sheep farmer, conservationist, naturalist, and businesswoman Becky Weed and her husband, David Tyler, raise sheep on pastures not treated with chemicals and guard their flocks with dogs and llamas, not guns. In the New West, they and others are pioneering fresh approaches to living in harmony with the wild.

ERIKA LARSEN

BIBLIOGRAPHY

Achenbach, Joel. 2009. "When Yellowstone Explodes." *National Geographic,* August.

Adams, Michael J., et al. 2013. "Trends in Amphibian Occupancy in the United States." *PLoS ONE,* 8(5).

Askins, Renée. 2002. *Shadow Mountain. A Memoir of Wolves, a Woman, and the Wild.* New York: Anchor Books/Random House.

Barber-Meyer, Shannon M. 2015. "Trophic Cascades From Wolves to Grizzly Bears or Changing Abundance of Bears and Alternate Foods?" *Journal of Animal Ecology,* 84.

Barber-Meyer, Shannon M., L. David Mech, and P. J. White. 2008. "Elk Calf Survival and Mortality Following Wolf Restoration to Yellowstone National Park." *Wildlife Monographs,* The Wildlife Society, 169.

Berger, Joel. 2004. "The Last Mile: How to Sustain Long-Distance Migration in Mammals." *Conservation Biology,* 18(2).

Bjornlie, Daniel D., et al. 2013. "Methods to Estimate Distribution and Range Extent of Grizzly Bears in the Greater Yellowstone Ecosystem." *Wildlife Society Bulletin,* DOI: 10.1002/wsb.368.

Bjornlie, Daniel D., et al. 2014. "Whitebark Pine, Population Density, and Home-Range Size of Grizzly Bears in the Greater Yellowstone Ecosystem." *PLoS ONE,* 9(2).

Black, George. 2012. *Empire of Shadows: The Epic Story of Yellowstone.* New York: St. Martin's Press.

Botkin, Daniel B. 1990. *Discordant Harmonies: A New Ecology for the Twenty-first Century.* New York, Oxford: Oxford University Press.

Brock, Thomas D. 1967. "Life at High Temperatures." *Science,* 158.

Brock, Thomas D. 1995. "The Road to Yellowstone— And Beyond." *Annual Review of Microbiology,* 49.

Brock, Thomas D., and Hudson Freeze. 1969. "*Thermus aquaticus* gen. n. and sp. n., a Non-sporulating Extreme Thermophile." *Journal of Bacteriology,* 98(1).

Cahill, Tim. 2004. *Lost in My Own Backyard: A Walk in Yellowstone National Park.* New York: Crown Publishers.

Chang, Wu-Lung, et al. 2007. "Accelerated Uplift and Magmatic Intrusion of the Yellowstone Caldera, 2004 to 2006." *Science,* 318.

Chase, Alston. 1987. *Playing God in Yellowstone: The Destruction of America's First National Park.* San Diego: Harcourt Brace Jovanovich.

Clark, Susan G. 2008. *Ensuring Greater Yellowstone's Future: Choices for Leaders and Citizens.* New Haven and London: Yale University Press.

Comstock, Theo. B. 1874. "The Yellowstone National Park." *The American Naturalist,* VIII(2).

Corn, Paul Stephen. 2007. "Amphibians and Disease: Implications for Conservation in the Greater Yellowstone Ecosystem." *USGS Staff—Published Research,* Paper 103.

Costello, Cecily M., et al. 2014. "Influence of Whitebark Pine Decline on Fall Habitat Use and Movements of Grizzly Bears in the Greater Yellowstone Ecosystem." *Ecology and Evolution,* 4(10).

Craighead, Frank C., Jr. 1979. *Track of the Grizzly.* San Francisco: Sierra Club Books.

Craighead, John J., Jay S. Sumner, and John A. Mitchell. 1995. *The Grizzly Bears of Yellowstone: Their Ecology in the Yellowstone Ecosystem, 1959–1992.* Washington, D.C.: Island Press.

D'Anastasio, R., et al. 2011. "Origin, Evolution, and Paleoepidemiology of Brucellosis." *Epidemiology and Infection,* 139(1).

Doak, Daniel F., and Kerry Cutler. 2013. "Re-evaluating Evidence for Past Population Trends and Predicted Dynamics of Yellowstone Grizzly Bears." *Conservation Letters,* 0(2013).

Doak, Daniel F., and Kerry Cutler. 2014. "Van Manen et al., Doth Protest Too Much: New Analyses of the Yellowstone Grizzly Population Confirm the Need to Reevaluate Past Population Trends." *Conservation Letters,* 7(3), 1-2.

Duffield, John W., Chris J. Neher, and David A. Patterson. 2008. "Wolf Recovery in Yellowstone: Park Visitor Attitudes, Expenditures, and Economic Impacts." *The George Wright Forum,* 25(1).

Eisenhauer, Nico, et al. 2007. "Invasion of a Deciduous Forest by Earthworms: Changes in Soil Chemistry, Microflora, Microarthropods and Vegetation." *Soil Biology and Biochemistry,* 39(5).

Ernst, Joseph W., editor. 1991. *Worthwhile Places: Correspondence of John D. Rockefeller, Jr. and Horace M. Albright.* New York: Fordham University Press.

Farrell, Jamie, et al. 2014. "Tomography From 26 Years of Seismicity Revealing That the Spatial Extent of the Yellowstone Crustal Magma Reservoir Extends Well Beyond the Yellowstone Caldera." *Geophysical Research Letters,* 41(9).

Farrell, Justin. 2015. *The Battle for Yellowstone: Morality and the Sacred Roots of Environmental Conflict.* Princeton, N.J.: Princeton University Press.

Fischer, Hank. 1995. *Wolf Wars: The Remarkable Inside Story of the Restoration of Wolves to Yellowstone.* Helena, Mont.: Falcon Press.

Fishbein, Seymour L. 1997. *Yellowstone Country.* National Geographic Park Profiles. Washington, D.C.: National Geographic.

Fortin, Jennifer K., et al. 2013. "Dietary Adjustability of Grizzly Bears and American Black Bears in Yellowstone National Park." *Journal of Wildlife Management,* 77(2).

Frelich, Lee E., et al. 2006. "Earthworm Invasion Into Previously Earthworm-Free Temperate and Boreal Forests." *Biological Invasions,* 8(6).

Glick, Dennis, Mary Carr, and Bert Harting, editors. 1991. *An Environmental Profile of the Greater Yellowstone Ecosystem.* Bozeman, Mont.: The Greater Yellowstone Coalition.

Gould, William R., et al. 2012. "Estimating Occupancy in Large Landscapes: Evaluation of Amphibian Monitoring in the Greater Yellowstone Ecosystem." *Wetlands,* 32.

Grinnell, George Bird. 1883. "The Park Monopolists Checked." *Forest and Stream,* 19(24).

Grinnell, Joseph, and Tracy I. Storer. 1916. "Animal Life as an Asset of National Parks." *Science,* 44(1133).

Gunther, Kerry A., et al. 2014. "Dietary Breadth of Grizzly Bears in the Greater Yellowstone Ecosystem." *Ursus,* 25(1).

Haines, Aubrey L. 1977. *The Yellowstone Story: A History of Our First National Park.* Volume One. Revised Edition. Niwot: University Press of Colorado.

Haines, Aubrey L. 1997. *The Yellowstone Story. A History of Our First National Park.* Volume Two. Revised Edition. Boulder: University Press of Colorado.

Hendrix, Paul F. 2006. "Biological Invasions Belowground—Earthworms as Invasive Species." *Biological Invasions,* 8.

Hossack, Blake R., et al. 2015. "Trends in Rocky Mountain Amphibians and the Role of Beaver as a Keystone Species." *Biological Conservation,* 187.

Houston, Douglas B. 1982. *The Northern Yellowstone Elk: Ecology and Management.* New York: Macmillan Publishing Co.

Huang, Hsin-Hua, et al. 2015. "The Yellowstone Magmatic System From the Mantle Plume to the Upper Crust." *Science,* 348(6236).

Hugenholtz, Philip, et al. 1998. "Novel Division Level Bacterial Diversity in a Yellowstone Hot Spring." *Journal of Bacteriology,* 180(2).

Hutton, Paul. 1985. *Phil Sheridan & His Army.* Norman: University of Oklahoma Press.

Hutton, Paul. 1985. "Phil Sheridan's Crusade for Yellowstone." *American History Illustrated,* 19(13).

Johnston, Jeremy. 2002. "Preserving the Beasts of Waste and Desolation: Theodore Roosevelt and Predator Control in Yellowstone." *Yellowstone Science,* (10)2.

Keiter, Robert B., and Mark S. Boyce, editors. 1991. *The Greater Yellowstone Ecosystem: Redefining America's Wilderness Heritage.* New Haven, Conn.: Yale University Press.

Koch, Edward D., and Charles R. Peterson. 1995. *Amphibians & Reptiles of Yellowstone and Grand Teton National Parks.* Salt Lake City: University of Utah Press.

Loendorf, Lawrence L., and Nancy Medaris Stone. 2006. *Mountain Spirit: The Sheep Eater Indians of Yellowstone.* Salt Lake City: University of Utah Press.

Logan, Jesse A., and Barbara J. Bentz. 1999. "Model Analysis of Mountain Pine Beetle (Coleoptera: Scolytidae) Seasonality." *Environmental Entomology,* 28(6).

Logan, Jesse A., and James A. Powell. 2001. "Ghost Forests, Global Warming, and the Mountain Pine Beetle (Coleoptera: Scolytidae)." *American Entomology,* Fall.

Logan, Jesse A., William W. MacFarlane, and Louisa Willcox. 2010. "Whitebark Pine Vulnerability to Climate-Driven Mountain Pine Beetle Disturbance in the Greater Yellowstone Ecosystem." *Ecological Applications,* 20(4).

MacArthur, Robert H., and Edward O. Wilson. 1967. *The Theory of Island Biogeography.* Princeton, N.J.: Princeton University Press.

Macfarlane, William W., Jesse A. Logan, and Wilson R. Kern. 2013. "An Innovative Aerial Assessment of Greater Yellowstone Ecosystem Mountain Pine Beetle-Caused Whitebark Pine Mortality." *Ecological Applications,* 23(2).

Magoc, Chris J. 1999. *Yellowstone: The Creation and Selling of an American Landscape, 1870–1903.* Albuquerque: University of New Mexico Press.

Marcus, W. Andrew, et al. *Atlas of Yellowstone.* 2012. Berkeley: University of California Press.

Marshall, Kristin N., N. Thompson Hobbs, and David J. Cooper. 2012. "Stream Hydrology Limits Recovery of Riparian Ecosystems After Wolf Reintroduction." *Proceedings of the Royal Society B,* 280(1756).

Massey, Jack, Sarah Cubaynes, and Tim Coulson. 2013. "Will Central Wyoming Elk Stop Migrating to Yellowstone, and Should We Care?" *Ecology,* 94(6).

Mattson, David J., Marilynn G. French, and Steven P. French. 2002. "Consumption of Earthworms by Yellowstone Grizzly Bears." *Ursus,* 13.

McMenamin, Sarah K., Elizabeth A. Hadly, and Christopher K. Wright. 2008. "Climatic Change and Wetland Desiccation Cause Amphibian Decline in Yellowstone National Park." *Proceedings of the National Academy of Sciences,* 105(44).

McMillion, Scott. 1998. *Mark of the Grizzly: True Stories of Recent Bear Attacks and the Hard Lessons Learned.* Helena, Mont.: Falcon Press.

McNamee, Thomas. 1997. *The Grizzly Bear.* New York: Lyons and Burford.

McNamee, Thomas. 2014. *The Killing of Wolf Number Ten: The True Story.* Westport, Conn.: Prospecta Press.

McNamee, Thomas. 1997. *The Return of the Wolf to Yellowstone.* New York: Henry Holt and Co.

Meagher, Margaret Mary. 1973. "The Bison of Yellowstone National Park." National Park Service Scientific Monograph Series, 1.

Meagher, Mary, and Margaret E. Meyer. 1994. "On the Origin of Brucellosis in Bison in Yellowstone National Park: A Review." *Conservation Biology,* 8(3).

Mernin, Jerry. Forthcoming. *Yellowstone Ranger. Stories From a Life Spent With Bears, Backcountry Horses, and Mules From Yosemite to Yellowstone.* Helena, Mont.: Riverbend Publishing.

Middleton, Arthur D. 2014. "Is the Wolf a Real American Hero?" *International New York Times,* March 9, A21.

Middleton, Arthur D., et al. 2013. "Animal Migration Amid Shifting Patterns of Phenology and Predation: Lessons From a Yellowstone Elk Herd." *Ecology,* 94(6).

Middleton, Arthur D., et al. 2013. "Grizzly Bear Predation Links the Loss of Native Trout to the Demography of Migratory Elk in Yellowstone." *Proceedings of the Royal Society B,* 280(1762).

Middleton, Arthur D., et al. 2013. "Rejoinder: Challenge and Opportunity in the Study of Ungulate Migration Amid Environmental Change." *Ecology,* 94(6).

Monahan, William B., Nicholas A. Fisichelli. 2014. "Climate Exposure of US National Parks in a New Era of Change." *PLoS ONE,* 9(7).

Munro, Andrew R., Thomas E. McMahon, and James R. Ruzycki. 2005. "Natural Chemical Markers Identify Source and Date of Introduction of an Exotic Species: Lake Trout (*Salvelinus namaycush*) in Yellowstone Lake." *Canadian Journal of Fisheries and Aquatic Sciences,* 62(1).

Murphy, Peter J., et al. 2009. "Distribution and Pathogenicity of *Batrachochytrium dendrobatidis* in Boreal Toads From the Grand Teton Area of Western Wyoming." *EcoHealth,* 6(1).

National Geographic. 2012. *Guide to National Parks of the United States.* Seventh Edition. Washington, D.C.: National Geographic.

National Park Service. 2010. "Amphibian Monitoring in the Greater Yellowstone Network—Project Report 2008 and 2009, Yellowstone and Grand Teton Parks." National Resource Data Series, 2010/072.

Nelson, Abigail A., et al. 2016. "Native Prey Distribution and Migration Mediates Wolf Population on Domestic Livestock in the Greater Yellowstone Ecosystem." Published on the Web February 10, 2016. *Canadian Journal of Zoology,* 10.1139/cjz-2015-0094.

Newmark, William D. 1987. "A Land-Bridge Island Perspective on Mammalian Extinctions in Western North American Parks." *Nature,* 325(6103).

Peacock, Doug. 2014. "What It Takes to Kill a Grizzly Bear." *The Daily Beast,* Available online at thedailybeast.com/articles/2014/11/23.

Pease, Craig M., and David J. Mattson. 1999. "Demography of the Yellowstone Grizzly Bears." *Ecology,* 80(3).

Persico, Lyman, and Grant Meyer. 2012. "Natural and Historical Variability in Fluvial Processes, Beaver Activity, and Climate in the Greater Yellowstone Ecosystem." Published on the Web November 29, 2012. *Earth Surface Processes and Landforms,* 10.1002/esp.3349.

Peterson, Rolf O., et al. 2014. "Trophic Cascades in a Multicausal World: Isle Royale and Yellowstone." *Annual Review of Ecology, Evolution, and Systematics,* 45.

Pritchard, James A. 1999. *Preserving Yellowstone's Natural Conditions: Science and the Perception of Nature.* Lincoln: University of Nebraska Press.

Raffa, Kenneth F., et al. 2008. "Cross-scale Drivers of Natural Disturbances Prone to Anthropogenic Amplification: The Dynamics of Bark Beetle Eruptions." *BioScience,* 58(6).

Ray, Andrew, et al. 2014. "Using Monitoring Data to Map Amphibian Breeding Hotspots and Describe Wetland Vulnerability in Yellowstone and Grand Teton National Parks." *Park Science,* 31(1).

Reese, Rick. 1991. *Greater Yellowstone: The National Park and Adjacent Wildlands.* Helena, Mont.: American and World Geographic Publishing.

Reiger, John F. 1975. *American Sportsmen and the Origins of Conservation.* New York: Winchester Press.

Reinhart, Daniel P., et al. 2001. "Effects of Exotic Species on Yellowstone's Grizzly Bears." *Western North American Naturalist,* 61(3).

Righter, Robert W. 1982. *Crucible for Conservation: The Struggle for Grand Teton National Park.* Moose, Wyo.: Grand Teton Natural History Association.

Ripple, William J., et al. 2014. "Trophic Cascades From Wolves to Grizzly Bears in Yellowstone." *Journal of Animal Ecology,* 83(1).

Robison, Hillary L. 2009. *Relationships Between Army Cutworm Moths and Grizzly Bear Conservation.* Dissertation, University of Nevada, Reno. Available online at gradworks.umi.com/33/87/3387820.html.

Romme, William H., et al. 2011. "Twenty Years After the 1988 Yellowstone Fires: Lessons About Disturbance and Ecosystems." *Ecosystems,* 14(7).

Runte, Alfred. 1987. *National Parks: The American Experience.* Lincoln: University of Nebraska Press.

Schullery, Paul. 1992. *The Bears of Yellowstone.* Worland, Wyo.: High Plains Publishing Co.

Schullery, Paul. 2010. "Greater Yellowstone Science: Past, Present, and Future." *Yellowstone Science,* 18(2).

Schullery, Paul. 2004. *Searching for Yellowstone: Ecology and Wonder in the Last Wilderness.* Helena: Montana Historical Society Press.

Schullery, Paul, editor. 2013. *Yellowstone Bear Tales: Adventures, Mishaps, and Discoveries Among the World's Most Famous Bears.* Second edition. Available online at paulschullery.com.

Schullery, Paul, and Whittlesey, Lee. 2003. *Myth and History in the Creation of Yellowstone National Park.* Lincoln: University of Nebraska Press.

Schwartz, Charles C., et al. 2013. "Body and Diet Composition of Sympatric Black and Grizzly Bears in the Greater Yellowstone Ecosystem." *The Journal of Wildlife Management,* DOI: 10.1002/jwmg.633.

Sellars, Richard West. 1997. *Preserving Nature in the National Parks: A History.* New Haven, Conn.: Yale University Press.

Shaffer, Mark Leslie. 1978. *Determining Minimum Viable Population Sizes: A Case Study of the Grizzly Bear (Ursus arctos L.).* Dissertation, Department of Forestry and Environmental Sciences, Duke University.

Smith, Douglas W. 2005. "Ten Years of Yellowstone Wolves. 1995–2005." *Yellowstone Science,* 13(1).

Smith, Douglas W., and Gary Ferguson. 2012. *Decade of the Wolf: Returning the Wild to Yellowstone, Revised and Updated.* Guilford, Conn.: Lyons Press.

Smith, Robert B., and Lee J. Siegel. 2000. *Windows Into the Earth: The Geologic Story of Yellowstone and Grand Teton National Parks.* Oxford: Oxford University Press.

Smith, Douglas W., and Daniel B. Tyers. 2012. "The History and Current Status and Distribution of Beavers in Yellowstone National Park." *Northwest Science,* 86(4).

Stahler, Daniel R., et al. 2013. "The Adaptive Value of Morphological, Behavioural and Life-History Traits in Reproductive Female Wolves." *Journal of Animal Ecology,* 82(1).

Suchy, Willie J., et al. 1985. "New Estimates of Minimum Viable Population Size for Grizzly Bears of the Yellowstone Ecosystem." *Wildlife Society Bulletin,* 13(3).

Talluto, Matthew V., and Craig W. Benkman. 2014. "Conflicting Selection From Fire and Seed Predation Drives Fine-Scaled Phenotypic Variation in a Widespread North American Conifer." *Proceedings of the National Academy of Sciences,* 111(26).

Teisberg, Justin E., et al. 2014 "Contrasting Past and Current Numbers of Bears Visiting Yellowstone Cutthroat Trout Streams." *The Journal of Wildlife Management,* DOI: 10.1002/jwmg.667.

Tiunov, Alexei V., et al. 2006. "Invasion Patterns of Lumbricidae Into the Previously Earthworm-Free Areas of Northeastern Europe and the Western Great Lakes Region of North America." *Biological Invasions,* DOI 10.1007/s10530-006-9018-4.

Van Manen, Frank T., et al. 2014. "Re-evaluation of Yellowstone Grizzly Bear Population Dynamics Not Supported by Empirical Data: Response to Doak & Cutler." *Conservation Letters,* 7(3).

White, P. J., Robert A. Garrott, and Glenn E. Plumb, editors. 2013. *Yellowstone's Wildlife in Transition.* Cambridge, Mass.: Harvard University Press.

White, P. J., et al. 2010. "Migration of Northern Yellowstone Elk: Implications of Spatial Structuring." *Journal of Mammalogy,* 91(4).

White, P. J., Rick L. Wallen, and David E. Hallac, editors. 2015. *Yellowstone Bison: Conserving an American Icon in Modern Society.* Yellowstone National Park: Yellowstone Association.

Wilkinson, Todd. 2013. *Last Stand: Ted Turner's Quest to Save a Troubled Planet.* Guilford, Conn.: Lyons Press.

Wolf, Evan C., David J. Cooper, and N. Thompson Hobbs. 2007. "Hydrologic Regime and Herbivory Stabilize an Alternative State in Yellowstone National Park." *Ecological Applications,* 17(6).

ACKNOWLEDGMENTS

This book wouldn't exist had not Chris Johns, then editor-in-chief of *National Geographic* magazine, conceived the idea of publishing a special issue of the magazine, to appear during the centennial year of the U.S. National Park Service, 2016, devoted entirely to the Greater Yellowstone Ecosystem. Chris summoned to the task an expert team of photographers, editors, mapmakers, graphic designers, and others; he also, to my surprise, gave me the responsibility and the honor of writing the main text for the entire issue. That became the May 2016 issue of *National Geographic,* published under the leadership of the present editor-in-chief, Susan Goldberg, to whom also I owe a vast debt of thanks. I've been in the magazine-writer guild for almost 40 years now, and I consider myself deeply fortunate to have worked with two top editors as dedicated, savvy, and likable as Chris and Susan.

I mentioned the team of National Geographic people mustered to the Yellowstone project (which was unprecedented in scope for that magazine, and I suppose almost for any magazine), and though I can't list them all by name, I want to call out at least some. Michael (Nick) Nichols, my friend and working partner from many previous National Geographic projects (going back to the Megatransect, in 1999–2000, our effort to document Mike Fay's footslog path across the central African forests), led a squad of photographers who worked throughout the ecosystem on the Yellowstone project; Nick helped guide and coordinate their efforts, while doing a large share of the shooting himself. Back in Washington, at what I think of as the Nat Geo mother ship, Kathy Moran was the photo editor to whom these hundreds of thousands of images flowed, and who bore first responsibility for winnowing the great from the merely excellent. The photographers, in addition to Nick Nichols, were these: Ronan Donovan, David Guttenfelder, Charlie Hamilton James, Louise Johns, Erika Larsen, Cory Richards, Joe Riis, and Drew Rush. Bill Marr, meanwhile, worked on the design side, from the earliest days of the effort, to bring our work all together in a felicitous shape. Jamie Shreeve, deputy editor of the magazine and my editor for the magazine text, pushed me hard to move beyond early drafts toward better structures and choices of expression and helped me toward those improvements with a lot of deft, smart, tireless editing. Todd Wilkinson, my colleague and friend here in Bozeman, Montana, with decades of journalistic experience covering the Greater Yellowstone Ecosystem, was called in to write the captions and most of the sidebars (not just for the magazine issue but here again in the book). In addition to these people there were many others at National Geographic, or closely associated with it, who made the special issue happen—too many to name, but including: Darlene and Jeff Anderson, Martin Gamache, David Lindsey, Karen Sligh, Kaitlin Yarnall, and the quartet of keen researchers who double-checked and sometimes corrected my facts: Christy Barcus, Julie Beer, Dave Lande, and Brad Scriber.

Lisa Thomas and Susan Tyler Hitchcock, of National Geographic Books, took over where the magazine work left off. Together with Sanaa Akkach, Laura Lakeway, and Judith Klein, they have guided this book expertly into the shape you now hold.

My wife, Betsy Gaines Quammen, bore a goodly burden in support of this work, both by way of running our household alone during my frequent absences and, maybe harder still, enduring my focused and sometimes cranky moods when I was present. Fortunately, she believes in the values of wild landscape and wild creatures as much as I do, so she was a cheerful, encouraging partner throughout.

On the other side of the enterprise from my institutional home, National Geographic, are all the dozens of people who consented to interviews, supplied me with information and insights, tolerated my company as they did their work in the backcountry, and otherwise accommodated my efforts in gathering the story.

At Yellowstone National Park, Superintendent Dan Wenk opened the door generously, even while carefully maintaining the regulatory strictures to which we were subject, and gave his wisdom as well as his trust to me and the photographers. Others of YNP, present or former public servants to whom I owe thanks of all sorts: Bob Barbee, Pat Bigelow, Norm Bishop, Gary Brown, Jennifer Carpenter, Steve Cook, Colleen Curry, Brian Ertel, Chris Geremia, Kerry Gunther, Hank Haesler, Dave Hallac, John Kerr, Todd Koel, Rick McIntyre, Eric Morey, Zara Osman, Staffan Peterson, Tim Reid, Ann Rodman, Tobin Roop, Paul Schullery, Stacey Sigler, Doug Smith, Dan Stahler, Rick Wallen, Tammy Wert, P. J. White, and Lee Whittlesey.

At Grand Teton National Park, Superintendent David Vela was likewise generously welcoming and forcefully helpful to my efforts. Among his people and others who work within the park, I was also aided by: Steve Cain, Sarah Dewey, Millie Jimenez, Chris Perkins, Gary Pollock, and Jackie Skaggs, among others.

Jon Jarvis, director of the National Park Service, not only sanctioned these interactions but also gave me his own thoughts on the mission of the NPS and the American landscapes it protects.

Throughout the Greater Yellowstone Ecosystem, there were many other people—agency biologists and administrators, conservation activists, university-based scientists, private lands owners, and other citizens—who gave me their time, their trust, their thoughts, their opinions, sometimes a glimpse of their passions and frustrations and concerns about this place, in its particulars and as a whole. Rather than sorting them by affiliation or role or geography or political stripe, I'm going to mix them all together in one long litany of gratitude. Why? Not just to save a few words here but because I hope the very mixing, mildly perverse as it may seem to some of them, might presage a trend in the future for us all to come together, listening mutually and acting cooperatively for the welfare of Yellowstone as an ecosystem and an ideal, the great bear that stands at the heart of it, and all its other living creatures and physical factors. So, thank you: Peter Aengst, Ben Alexander, Joe Alexander, Hannibal Anderson, Lenox Baker, Curt Bales, Joel Berger, Dan Bjornlie, Mitch Bock, Tom Brock, Claudeo Broncho, Mark Bruscino, Caroline Byrd, Franz Camenzind, Scott Christensen, Susan Clark, Eric Cole, Aly Courtemanche, Lance Craighead, Marna Daley, Arny Dood,

Mike Ebinger, Bob Ekey, Luke Ellsbury, Mary Erickson, Joe Fidel, Kevin Frey, Mark Gocke, Andy Hanson, Mark Haroldson, Jodi Hilty, Hal Hunter, Lori Iverson, Joe Josephson, Virginia Kelly, Ted Kerasote, Bianca Klein, Mark Kossler, Karen Kress, Lee Livingston, Wes Livingston, Jesse Logan, Blake Lowrey, Doug McWhirter, Sava Malachowski, Dave Mattson, Cindy Mernin, Arthur Middleton, Bob Model, Peter Moyer, Abby Nelson, Kirk Nordstrom, Deb Patla, Doug Peacock, Mike Phillips, Andy Pils, Jim Pope, Chuck and Penny Preston, Steve Primm, Elli Radinger, Ray Rasker, Andy Ray, Alan Redfield, Rick Reese, Frank Rigler, Hillary Robison, Stephanie Seay, Chris Servheen, Mark Shelton, Senator Alan and Ann Simpson, Bob Smith, Roger Stradley, Dan Thompson, Dan Tyers, Frank van Manen, Nathan Varley, Dan Vermillion, Nichole Walker, Pete Walsh, Jeff Welsch, Michael Whitefield, Louisa Wilcox, Ben Wise, and Mark Young.

Doubtless I've omitted some names that should be included. I live in this ecosystem and have for three decades, after all, and wherever I go, good people of very diverse interests and attitudes help me understand it better. Thank you. Thanks also to the grizzly bears, who teach what wildness is more vividly than anybody.

—D. Q.

ILLUSTRATIONS CREDITS

National Geographic Magazine Yellowstone Contributing Photographers

Michael Nichols: 2-3, 4-5, 10, 12, 16-17, 18, 42, 43, 46 and 47 (both with Ronan Donovan and the National Park Service), 52-9, 63, 64-5, 67, 69, 70-71, 86, 132, 137, 152, 153, 172, 180, 181, 184-5, 192-3, 204-205, 212-13; **Charlie Hamilton James:** 6-7, 8-9, 62 (UP), 68, 76-7, 91, 98-9, 100, 101, 106-107, 110, 111, 114-19, 124, 139, 140-141, 142, 143 ; **David Guttenfelder**: 34, 72, 83, 109, 150, 186, 187, 188-9, 194, 195; **Joe Riis:** 44-5, 78, 122-3, 144-5, 154-166, 198-9, 206-207; **Erika Larsen:** 49, 75, 121, 169, 182-3, 191, 208-209, 215; **Ronan Donovan:** 97, 104-105, 112-13, 129, 146-9, 151; **Drew Rush:** 108, 167 (with National Park Service); **Cory Richards:** 170-171; **Louise Johns:** 196, 197

Additional Photographs
23, Courtesy National Park Service; 26, Montana Historical Society Research Center, H-3614; 31, Jimmy Chin; 38-9, Raul Touzon/National Geographic Creative; 40-41, Barrett Hedges/National Geographic Creative; 50-51, Gordon Wiltsie/National Geographic Creative; 60, Underwood & Underwood/The Print Collector/Getty Images; 61 (UP-ALL), Yellowstone National Park Photo Collection; 61 (LO LE), Corbis; 61 (LO RT), Poster by Ludwig Hohlwein; Photo from Swim Ink 2, LLC/Corbis; 62 (LO), Eric Kruszewski/National Geographic Creative; 66, Robb Kendrick; 73, Jonathan Blair; 94, Bob Cochran/National Geographic Your Shot; 102, Barrett Hedges/National Geographic Creative; 103, Michael Forsberg/National Geographic Creative; 177, Tom Murphy/National Geographic Creative; 200, W. H. Jackson/The New-York Historical Society/Getty Images; 201 (UP LE), "The Grand Canyon of the Yellowstone" by Thomas Moran—Photo by Geoffrey Clements/Corbis; 201 (UP RT, LO LE, LO RT), Yellowstone National Park Photo Collection; 202-203, Tom Murphy/National Geographic Creative; 210-11, Tim Fitzharris/Minden Pictures.

Map Credits
14, USGS; Yellowstone National Park
15, Bureau of Land Management; Andrew J. Hansen, Montana State University; Montana Department of Revenue; National Conservation Easement Database; The Nature Conservancy; USGS

ABOUT THE AUTHORS

David Quammen is an author and journalist whose 15 books include *The Song of the Dodo* (1996), *The Reluctant Mr. Darwin* (2006), and *Spillover* (2012). Quammen is a contributing writer for *National Geographic,* in whose service he travels often, usually to wild places. In 2012 he received the Stephen Jay Gould Prize from the Society for the Study of Evolution. He lives in Bozeman, Montana, with his wife, Betsy Gaines Quammen, a conservationist at work on a doctorate in environmental history, and their family of other mammals.

Todd Wilkinson, author of the captions and sidebars in this book, has been writing about the environment for 30 years on assignments that have taken him around the world. His work has appeared in *National Geographic,* the *Washington Post,* the *Christian Science Monitor, Audubon, Scientific American,* the *Wall Street Journal,* and many other publications. He is the author of several books including *Last Stand: Ted Turner's Quest to Save a Troubled Planet.* He lives in Bozeman, Montana.

Published by National Geographic Partners
1145 17th Street NW, Washington, DC 20036

Library of Congress Cataloging-in-Publication Data
Names: Quammen, David, 1948-
Title: Yellowstone : a journey through America's wild heart / David Quammen.
Description: Washington, D.C. : National Geographic, [2016] | Includes
   bibliographical references.
Identifiers: LCCN 2016014893 | ISBN 9781426217548
Subjects: LCSH: Yellowstone National Park. | National parks and reserves--Wyoming.
   | National parks and reserves--Idaho. | National parks and reserves--Montana. | Nature
   conservation--Social aspects--Yellowstone National Park. | Outdoor recreation--Social
   aspects--Yellowstone National Park. | Public spaces--Social aspects--Yellowstone National Park.
Classification: LCC SB482.W8 Q83 2016 | DDC 333.7509787/52--dc23
LC record available at https://urldefense.proofpoint.com/v2/url?u=https-3A__lccn.loc.gov_201-
6014893&d=CwIFAg&c=uw6TLu4hwhHdiGJOgwcWD4AjKQx6zvFcGEsbfiY9-EI&r=NzwTuDLIyA2L
24ee5BSEjNZ3hB9PzdA5RJ1PWfNuZAE&m=Frr4S-qNSXKiWLwkAeS4pTWzRJF1VKg8XrklB3mK13
c&s=n2HoraCZfJGmxUsg0jnL_DOMfWVUDWMldPyM3ptwpoA&e=

Since 1888, the National Geographic Society has funded more than 12,000 research, exploration, and preservation projects around the world. National Geographic Partners distributes a portion of the funds it receives from your purchase to National Geographic Society to support programs including the conservation of animals and their habitats.

National Geographic Partners
1145 17th Street NW
Washington, DC 20036-4688 USA

Become a member of National Geographic and activate your benefits today at natgeo.com/jointoday.

For information about special discounts for bulk purchases, please contact National Geographic Books Special Sales: specialsales@natgeo.com

For rights or permissions inquiries, please contact National Geographic Books Subsidiary Rights: bookrights@natgeo.com

Interior design: Sanaa Akkach / Bill Marr
Printed in the United States of America
16/WOR/3

YELLOWSTONE
PARK FOUNDATION
EST. 1996

For nearly 150 years Yellowstone National Park's exceptional beauty has inspired philanthropy among generous individuals, corporations, and foundations who contribute to its long-term preservation.

The Yellowstone Park Foundation, the park's official fund-raising partner, plays a vital role in expanding this tradition of philanthropic support. It works closely with the National Park Service to identify Yellowstone National Park's immediate priority needs and long-term funding challenges. Foundation staff members connect generous donors with opportunities to support these priorities and participate in park stewardship. The Foundation has raised over $100 million for over 325 significant Park priority projects.

The following six strategic initiatives encompass the areas where philanthropy can have a lasting and significant impact in the Park:

Visitor Experience

Every trip to Yellowstone should be magical, but heavy annual visitation can take its toll on trails, campgrounds, and other facilities. It can also put a strain on the Park's ability to provide educational opportunities. The Yellowstone Park Foundation supports projects that enhance recreation, safety, and accessibility, while bolstering both in-park and online education.

Wildlife, Wonders, and Wilderness

Yellowstone is home to the largest concentration of wildlife in the lower 48 states and has more geysers and hot springs than the rest of the world combined. The Yellowstone Park Foundation supports research and conservation projects to preserve the wildlife and other precious natural resources for which Yellowstone is famous.

Cultural Treasures

Yellowstone—the world's very first national park—is the keeper of stories. The Yellowstone Park Foundation supports projects that protect, preserve, research, and share information about Yellowstone's human past. Projects include support for the Park's museum collection, archaeological surveys, historic preservation, and more.

Ranger Heritage

Yellowstone rangers have no small job. They are charged with protecting the Park's 2.2 million acres of natural resources as well as the safety of visitors. They need trustworthy equipment, modern technology, reliable transportation, and suitable facilities. The Yellowstone Park Foundation supports projects that promote the effectiveness, safety, and efficiency of rangers.

Tomorrow's Stewards

Yellowstone is one of the world's premier outdoor classrooms and offers several award-winning, ranger-led educational programs for children. Yellowstone Park Foundation's support helps expand the reach of these programs to promote the appreciation and stewardship of Yellowstone in the next generation.

Greenest Park

Yellowstone has long been a leader in natural resource management but still uses large quantities of fossil fuel and treated water and generates much solid waste in serving millions of annual visitors. The Yellowstone Park Foundation supports projects that aim to reduce Yellowstone's ecological footprint and better preserve environmental resources.

Yellowstone Park Foundation
www.ypf.org
406-586-6303